SpringerBriefs in Environmental Science

SpringerBriefs in Environmental Science present concise summaries of cutting-edge research and practical applications across a wide spectrum of environmental fields, with fast turnaround time to publication. Featuring compact volumes of 50 to 125 pages, the series covers a range of content from professional to academic. Monographs of new material are considered for the SpringerBriefs in Environmental Science series.

Typical topics might include: a timely report of state-of-the-art analytical techniques, a bridge between new research results, as published in journal articles and a contextual literature review, a snapshot of a hot or emerging topic, an in-depth case study or technical example, a presentation of core concepts that students must understand in order to make independent contributions, best practices or protocols to be followed, a series of short case studies/debates highlighting a specific angle.

SpringerBriefs in Environmental Science allow authors to present their ideas and readers to absorb them with minimal time investment. Both solicited and unsolicited manuscripts are considered for publication.

More information about this series at http://www.springer.com/series/8868

Robert E. Beazley • James P. Lassoie

Himalayan Mobilities

An Exploration of the Impacts
of Expanding Rural Road Networks
on Social and Ecological Systems
in the Nepalese Himalaya

 Springer

Robert E. Beazley
Cornell University
Ithaca, NY, USA

James P. Lassoie
Cornell University
Ithaca, NY, USA

ISSN 2191-5547   ISSN 2191-5555   (electronic)
SpringerBriefs in Environmental Science
ISBN 978-3-319-55755-7   ISBN 978-3-319-55757-1   (eBook)
DOI 10.1007/978-3-319-55757-1

Library of Congress Control Number: 2017936947

Printed on acid-free paper

This Springer imprint is published by Springer Nature
The registered company is Springer International Publishing AG
The registered company address is: Gewerbestrasse 11, 6330 Cham, Switzerland

Preface

Roads are widely recognized by governments, aid institutions, and development planners as essential for economic development and poverty reduction in developing countries (e.g., see Van de Walle 2008; WB 2006; Howe and Richards 1984). Roads are the essential building blocks of development. Without roads, there can be no hydro project infrastructure such as dams and no electricity, no telephone towers, and limited access to advanced life support and health care institutions, quality education, communication networks, and a host of other critical social and economic services. Roads are essential for connecting local producers to markets, reducing transportation costs, and helping reduce the price of goods in formerly remote areas.

However, roads also can have negative influences that are far ranging, often causing unintended and unanticipated impacts on environmental, socioeconomic, and sociocultural spheres. Road construction can cause landslides, debris flows, and other forms of mass wasting as well as the air, water, light, and noise pollution associated with the vehicles that use them (e.g., see Evink 2002; Forman et al. 2002; Oxley et al. 1947). Land value and land use change can have differential socioeconomic impacts on local people both in the vicinity of the road and at varying scales away from the road (e.g., see Blaikie et al. 2002; Van de Walle 2008; WB 2006). In addition the increase in access to an area that a road brings can have substantial impacts on cultural traditions and practices (e.g., see Beazley 2013; Price 1989; Cruikshank 1985).

In Nepal, where there were virtually no roads until the 1950s, it is safe to say that every isolated community in the mountains wants a road (NTNC 2008). Roads are the number one infrastructure development project prioritized by Village Development Committees (VDCs) (UNDP 2011). Nepal has made substantial progress in expanding its road network including connecting 73 of its 75 district headquarters by road in recent years. However, there are still vast areas in the mountains with no road access. In Nepal, 23% of the population (~6 million people) live more than 4 h walk to the nearest road, 60% live more than 2 h away, and another 23% live greater than 4 days away. Some villages are as far as a 13-days walk from the nearest road. Children often walk 2–3 h to go to school and it is not unusual for a pregnant woman to walk 2 days to a health clinic (SDC 2008).

There is no denying the fact that such isolated areas need roads. Roads are life-savers for Nepalis because they ease the burden of carrying goods by human porters, increase the opportunities to find work and education outside the local area, and reduce the time it takes to get health care. Road construction also provides direct employment through the nongovernmental organization (NGO) sponsored Green Roads Program and the World Health Organization's Food for Work Program that pays locals in cash or food to build roads with manual labor. For example, the Swiss Agency for Cooperation and Development (SDC) is one of many NGOs in Nepal involved with rural road construction, and in collaboration with the Government of Nepal it employs more than 4000 Nepalis in their rural road construction program (SDC 2008). At the same time there needs to be a recognition that the way in which these roads are planned and built, whether they are *Green Roads,*[1] local roads, or central government roads, has a huge impact on the communities and the environment which they traverse. In the past road projects were often conceived of as connecting two locations with little consideration of what lay in between them, specifically what effects the road alignment might have not only on the immediate areas impacted but also on communities at varying distances away. There will always be both positive and negative impacts from road building, but with better understanding of how roads impact complex socio-ecological (i.e., coupled social and ecological) systems the chances of mitigating the negative impacts and enhancing the positive impacts are much greater.

Our orientation is geared toward providing a better understanding of the coupled social and ecological system of mobility and how that functions within the realm of road construction and mobility in the Nepalese Himalayas by reviewing information about road impacts and integrating it with our own research on the Kali Ghandaki and Marsyangdi Highways and the Annapurna Circuit Trail in Nepal (Beazley 2013), and the senior author's most recent research (2014–15) along the Trans Himalayan Highway (Rasuwa District). This book is organized under four general thematic categories. In Part I, Roads and Transportation, we summarize and update information about the history, distribution, functions, and impacts of road development, both globally and at Nepal's national level (Chap. 1). Then in Part II, Mobility, we examine what is now termed the *Mobilities Turn* as a basis for understanding the current academic context within which rural roads form a significant mobilities constituent (Chap. 2). This is followed by Part III, Challenges and Impacts of Building Roads in the Himalayas, where we provide an in-depth assessment, using numerous case studies including our own empirical research (Beazley 2013), of the positive and negative impacts of roads on the environmental (Chap. 3), socioeconomic (Chap. 4), and sociocultural (Chap. 5) spheres, and how they are interrelated and influence each other. Finally in Part IV, The Way Forward, we synthesize these findings by considering the future of Himalayan Mobilities in Nepal through the lens of recent events in Nepal including: (1) the 2015 earthquakes

---

[1] Decentralized local participatory environmentally friendly construction methods using simple technology that maximizes local resources, builds local capacity, and often provides employment for neighboring communities (Mulmi 2009).

(April–May); (2) the promulgation of the new Nepali constitution (September 2015); (3) the reaction to the constitution, which led to a 4-month blockade of the southern border (September 2015–January 2016) cutting off petroleum shipments from Nepal's only supplier, India, and Nepal's attempts to open new avenues for the purchase and transport of petroleum from China; and (4) the frequently changing governmental structure (Chap. 6). Through this analysis we hope to provide a useful platform for policy makers and planners for understanding the coupled social and ecological system of mobilities and how to enhance Himalayan Mobilities in the future.

References

Beazley, R. E. (2013). *Impacts of expanding rural road networks on communities in the Annapurna Conservation Area, Nepal*. M.S. Thesis, Department of Natural Resources, Cornell University, Ithaca, NY.

Blaikie, P., Cameron, J., & Seddon, D. (2002). Understanding 20 years of change in West-Central Nepal: Continuity and change in lives and ideas. *World Development, 30*(7), 1255–1270.

Cruikshank, J. (1985). The gravel magnet: Some social impacts of the Alaska Highway on Yukon Indians. In K. Coates (Ed.), *The Alaska Highway: Papers of the 40th Anniversary Symposium* (pp. 172–187). Vancouver: University of British Columbia Press.

Evink, G. L. (2002). *Interaction between roadways and wildlife ecology: A synthesis of highway practice. NCHRP Synthesis 305*. Washington, DC: Transportation Research Board—The National Academies.

Forman, R., Sperling, D., Bissonette, J., Clevenger, A., Cutshall, D., Fahrig, L., France, R., Goldman, C., Heanue, K., Jones, J., Swanson, F., Turrentine, T., & Winter, T. (2002). *Road ecology*. Washington, DC: Island Press.

Howe, J., & Richards, P. (Eds.). (1984). *Rural roads and poverty alleviation*. Boulder, CO: Westview Press.

Mulmi, A. (2009). Green road approach in rural road construction for the sustainable development of Nepal. *Journal of Sustainable Development, 2*(3), 149.

National Trust for Nature Conservation (NTNC). (2008). *Sustainable development plan of Mustang* (2008–2013). Kathmandu, Nepal: NTNC/GoN/UNEP. Retrieved from http://www.rrcap.unep. org/nsds/uploadedfiles/file/sa/np/mnmt/document/sd_masterplan_Mustang.pdf.

Oxley, D. J., Fenton, M. B., & Carmody, G. R. (1974). The effects of roads on populations of small mammals. *Journal of Applied Ecology, 11*, 51–59.

Price, D. P. (1989). *Before the bulldozer: The Nambiquara Indians and the World Bank*. Cabin John, MD: Seven Locks Press.

Swiss Agency for Development and Cooperation (SDC). (2008). *Nepal—Roads to prosperity partnership results*. Asia Brief, Nepal Rural Roads. Retrieved from http://www.swiss-cooperation. admin.ch/nepal/en/Home/Document_Archive.

United Nations Development Project (UNDP). (2011). *Economic analysis of local government investment in rural roads in Nepal 2011*. Kathmandu: UNDP. Retrieved from http://www. lgcdp.gov.np/home/pdf/RuralRoadReport.pdf.

Van de Walle, D. (2008). *Impact evaluation of rural roads project*. World Bank Report. Retrieved from http://siteresources.worldbank.org/INTISPMA/Resources/383704-1146752240884/ Doing_ie_series_12.pdf.

World Bank (WB). (2006). *Infrastructure: Lessons from the last two decades of World Bank Engagement*. Discussion Paper. Retrieved from http://www-wds.worldbank.org.

Acknowledgements

This book would not have been possible without the generous support of numerous individuals at Cornell and in Nepal. We received invaluable support and input from many graduate students and faculty at Cornell to whom we owe a debt of gratitude. In particular, we would like to thank Professors Kareem Aly-Kassam, Rebecca Schneider, and Stephen Morreale for their support, generosity with their time, and keen insight at crucial periods during this project. A special thanks also goes to Professors Kathryn and David March and all the staff at the Cornell Nepal Study Program (CNSP) who supported me in this research and provided invaluable help and ready advise. Furthermore, CNSP was instrumental in providing all the necessary visas, letters of recommendation, and support that we needed to conduct field research in Nepal. In addition, the senior author benefitted greatly from the excellent Nepali language courses he took at Cornell University with Mr. Shambhu Oja and Ms. Banu Oja and at CNSP. Their patient and friendly guidance helped him progress from beginner to advanced level and provided both authors with a much broader understanding of Nepali culture. The Fulbright Hays Doctoral Dissertation Abroad Fellowship provided very timely financial support and assistance and the Nepal Fulbright Program was instrumental in providing visas and support during the 2014–2015 field season. A big thanks goes to Laurie Vasily and her staff at the Nepal Fulbright Office for all their assistance. I owe a special thanks to Professor Bhanu Timseena (Associate Professor, Tribhuvan University, Kathmandu), who was very generous with his time in advising me on field methods and helping me train my field assistants (2014–2014).

The senior author's most recent (2014–2015) research in Nepal benefitted significantly through field trips and late night discussions with fellow colleagues Austin Lord (Cornell University), Galen Murton (James Madison University), and Matthäus Rest (Ludwig-Maximilians-Universität München).

Lastly, REB's field assistants and interpreters during his field trips, Bijay Chettri and Naryan Khuikal (2009–2010) and Sudan Bhattari, Kedar Bhattarai, Teg Thing, and Lhakpa Tamang (2014–2015), were indispensable while conducting the empirical research presented in book.

Contents

Photo 1 A mother with child making gravel for road construction by hammering larger rocks to make smaller rocks along the Marsyangdi Highway. This is a common sight along many road construction alignments in Nepal (R.E. Beazley 2019).

Photo 2 An Upper Mustang horse owner is confronted by a front-end loader widening the trail to make it a road. Expanding road networks in Upper Mustang have lead to the end of the centuries old horse culture in Upper Mustang as well as the horse trade between Nepal and Tibet. Now Mustangis would rather buy cheap Chinese motorcycles smuggled across the border than use their traditional means of travel-the horse. As one horse owner put it: "You only have to feed a motorcycle when you use it- a horse you have to feed and take care of whether you use it or not" (R.E. Beazley 2014).

About the Authors

Robert E. Beazley is a Ph.D. candidate in the Department of Natural Resources, Cornell University, USA. His M.S. research focused on the impacts of expanding rural road networks in the Nepal Himalayas. His current research involves an investigation of gendered mobility and borderland infrastructure in Rasuwa district, Nepal. E-mail: reb265@cornell.edu. Mailing address: Fernow Hall, Room 309, Cornell University, Ithaca, NY 14853, USA.

James P. Lassoie is a Professor in the Department of Natural Resources and an International Professor of Conservation in the College of Agriculture and Life Sciences at Cornell University. Originally trained as a forest ecologist at the University of Washington in Seattle, he now focuses on community-based conservation science and management and has worked extensively in Africa, Asia, Latin America, and North America. He currently holds a Professor appointment in the State Key Laboratory of Seedling Bioengineering, Ningxia Forest Institute, Yinchuan, Ningxia, and has worked in China since 1999. Professor Lassoie has over 150 scholarly research publications and has directed the development of a web-based educational system, www.conservationbridge.org, since 2007. E-mail: jpl4@cornell.edu. Mailing address: Fernow Hall, Room 201, Cornell University, Ithaca, NY 14853, USA.

Part I
Roads and Transportation

Chapter 1
A Global Review of Road Development

Abstract Historically roads began as an extension of trails adapted to handle increased traffic and wheeled vehicles. Many modern paved roads follow traditional trade routes such as the Silk Road, the Grand Trunk Road, and the Amber Road. Roads have profound impacts on humans, animals, and ecosystems—they are a prime example of what is called human environment interactions and/or coupled social and ecological systems. Roads can have both positive and negative influences that have far ranging, unanticipated, and unintended consequences on environmental, socioeconomic, and sociocultural spheres. When one considers the scale of impacts in all three spheres that travel beyond the areas immediately adjacent to the road surface, global road building must be acknowledged as one of the most profound and influential human activities on earth. Before 1950 motorable roads in Nepal were virtually nonexistent. It was not until 1956 that Kathmandu had a road usable by trucks that linked to the existing road in the flat Terai area connecting Kathmandu to the railhead and India—for almost 30 years, from 1927 until the completion of this section of road, the only way to transport goods into or out of the Kathmandu Valley was by the ropeway, human porter, or animal. Latest available figures suggest that as of 2016 Nepal has a total road length greater than 81,000 km. Recent events including the 2015 Gorkha Earthquake, the promulgation of the new Nepali constitution (September 2015) and the subsequent southern border blockade, and the continuing instability of the government all have an effect on people's ability to access and use these roads.

Keywords Road history • Trade routes • Human environment interactions • Coupled human and ecological systems • Nepal • Gorkha Earthquake • Southern blockade • New Nepali constitution

When we speak of roads in developed countries, we are usually referring to a tract that is specifically designed and developed for use by mechanized vehicles, mainly cars and trucks. However, this specific usage is a relatively new phenomena based in the Industrial Revolution. Before this period, a road referred to any corridor used for movement of humans and animals. In essence, roads are trails that have evolved to accommodate larger numbers of people, animals, and wheeled, often motorized, vehicles (Lay 1992). Many of our modern roads are built on top of ancient traditional trade routes, such as the Amber Road in Europe, the Grand Trunk Road in India, and the Silk Roads in Asia (Table 1.1). Roads are often referred to in the

© The Author(s) 2017   3
R.E. Beazley, J.P. Lassoie, *Himalayan Mobilities*, SpringerBriefs
in Environmental Science, DOI 10.1007/978-3-319-55757-1_1

development literature as essential in connecting previously "remote inaccessible" areas to modern urban centers and markets to help stimulate economic development and reduce poverty. However, many of these "remote inaccessible" areas were in fact on major trade routes and in relative terms experienced as much traffic in the past as our modern super highways do today. These trade routes acted as the Internet of their day spreading innovations, technologies, languages, spiritual teachings, military tactics and munitions, and other information from the origins of the diverse travelers that traversed them to far flung locations along the routes and then further by locals who travelled spur routes into other areas. In this sense, many of the types of changes that modern roads bring to these areas are not new, however the volume of traffic and the speed of change has increased dramatically. To those of us living in the developed world these dramatic changes were experienced a century ago. The thought of living in a village 2 weeks walk from a road head or even a 20-min walk away for us is unimaginable; we take roads for granted.

1.1   A Historical Perspective of Road Development

The history of roads predates the invention of the wheel. Roads or tracks developed as humans and animals traversed routes that led to food, water, shelter, and other humans and animals. As certain routes became more heavily used they were widened and improved making them more permanent tracks. Some became major trade routes and in many cases these eventually became modern roads. A few of the better-known routes are listed in Table 1.1.

Table 1.1 Trade routes that are now part of modern roads

Trade route	Section	Countries	Modern roads
Asia			
Silk Road	Lhasa to Kathmandu	Tibet, China & Nepal	China: Hwy 318 & Nepal: Arniko Hwy
Silk Road	Lhasa to Kathmandu	Tibet, China & Nepal	China: Hwy G219 & Nepal Trans Himalayan Hwy
Silk Road	Khasgar to Abbottabad	Xinjiang, China & Pakistan	Karakoram Hwy = Pakistan: Hwy N35 & China: Hwy G314
Silk Road-Grand Trunk Road	Abbottabad to Hassanabdal	Pakistan	Pakistan: Hwy N35 & Hwy N5
Grand Trunk Road	Chittagong to Delhi to Lahore to Kabul	Bangladesh, India, Pakistan, & Afghanistan	India: Hwy NH2, NH91, & NH1 & Golden Quadrilateral Project; Pakistan: Hwy N5; Afghanistan: Jalalabad-Kabul Road
Europe			
Amber Road	North Sea to Baltic Sea to Mediterranean Sea	Russia, Estonia, Lithuania, Poland, Czech Republic, Slovak Republic, Hungary, Austria, Slovenia, & Italy	Hungarian Road 86, E65

Early transport of goods was done either on the backs of animals or humans. As humans began using animals as beasts of burden the tracks became more hard packed and permanent. Some of the earliest examples of these types of permanent tracks date from around 6000 BC. The earliest evidence of constructed roads in the form of streets paved with stones dates to around 4000 BC in Mesopotamia near Ur and Babylon in what is now Iraq. The construction of these roads involved the use of mud bricks held in place with bitumen. In England corduroy roads built of logs and sand to facilitate travel through swampy areas date to the thirty-third century BC. Road building around the same time period in the Indian subcontinent and Indus Valley consisted of laying burned bricks bound with bitumen to form a solid enduring track (Lay 1992). Chandragupta (340–298 BC) founder of the Maurya Empire initiated a portion of what would become the Grand Trunk Road, eventually stretching from Chittagong, Bangladesh through India to Lahore, Pakistan and on to Kabul, Afghanistan. Around 3500 BC in Persia road building was progressing along a route that stretched more than 2400 km between the Mediterranean Sea and the Persian Gulf, making it one of the earliest long distance roads. Sections of this road became part of the Royal Roads of Darius I and Cyrus II in the sixth to fifth centuries BC, with two main stretches covering nearly 2250 km each (Lay 1992; Needham et al. 1954). In 2500 BC Pharaoh Cheops constructed another early road to facilitate building the Great Pyramid. One of the oldest paved roads still in existence was built around 2000 BC on the island of Crete. The Minoans connected the south coast of Crete over the mountains to the north coast by constructing a 50 km road of layered stone with sophisticated engineering techniques including the use of a clay-gypsum mortar to bind the stones, a raised center to drain water, and gutters to carry the drain water away (Lay 1992).

Meanwhile, in Europe by the mid sixteenth century BC four major trading routes intersected to make up what came to be known as the Amber Road due to the sources of amber in the Baltic Sea and its importance as a highly valued medium of exchange for salt and other trade goods (Hammarlund 2010; Schreiber 1961). By the fourth century BC the Romans had begun what would become the most methodical road network system of the ancient world, eventually extending more than 400,000 km of road throughout the Empire (Grant 1978). While initially these roads were mainly hardened earth or compacted gravel, as Roman road building technology advanced more sophisticated forms of paving to preserve the longevity of the roads were introduced. This often consisted of several different layers of stones with the top layer, a mixture sand and/or gravel, covered by large blocks of limestone or basalt from six to eight inches thick making the total depth of the road from three to six feet. This system of roads was instrumental in the rise of power of the Roman state and in shaping its economy, geography, and identity (Laurence 1999).

Early Chinese road networks include the Imperial Highway and the Great Wall as well as several major international trade route networks including the Silk Roads and the Tea Horse Trade route. The Imperial Highway saw major development during the Warring States period (475–221 BC) and by Qin Dynasty Emperor Shihuangdi during the third century BC. The main material of construction was 'water-bound macadam' of tamped rubble and gravel to form the surface of the roads (Needham et al. 1954), although Lay (1992: 79) claims the use of this term 'water-bound macadam' is inaccurate. Needham et al. (1954) make some interesting comparisons between Chinese and Roman road building as certain periods overlap, for example the construction of the Roman Appian Way and the Roman Appian Way and the Chinese military highway through the Chinn-ling Shan, which both began in 312 BC. In terms of construction in general, the Roman roads were deeper and more heavily constructed, while the lighter more elastic Chinese roads proved more durable over time. The total length of road networks of the Qin and Han Dynasties was approximately 35,406 km, whereas the Roman road networks exceeded 78,050 km of paved road and 400,000 km total road length (paved and unpaved). However, Needham et al. (1954) note that the Chinese road network was methodically integrated with their network of canals and waterways perhaps requiring less overland routes.

In South America the Incan road network stretched from Quito, Ecuador through Cuzco, Peru to Santiago, Chile. There were two main routes with many intersecting lateral trunk routes and regional and local offshoots connecting the highlands and the coast. The high route following the Andes (~5470 km) was challenging road-building requiring ingenious engineering to cross numerous rivers and passes higher than 5000 m. The second route following the coast (~3620 km) posed its own challenges with just as many river crossings. The total length of the combined network that made up the royal highway is estimated to be 30,000–40,000 km (Moseley 2001; Covey 2006).

Until the mid eighteenth century AD road building in Europe followed the same basic techniques established by the Roman road builders. However the increased importance of travel for expanding economies and military movement combined with improved methods of travel including better horse breeding and more efficient vehicles led to the invention of better road building techniques. Several people were important innovators in improving road-surfacing techniques including Tésaguet in France and Telford in Scotland but it was arguably John Loudon McAdam, also from Scotland, that had the most enduring influence. McAdam discovered that the use of small angular stones as a layer on a well-drained surface eliminated the need for an underlying rock layer as the angular stones when compacted interlocked creating their own cohesive force. To make it impenetrable this initial layer was covered by another layer of small stones. This technique, called 'macadam' after its inventor, became wide spread due to its success and is still in use today (Lay 1992).

1.2   Roads as Coupled Social and Ecological Systems

New understanding of the complex interconnected links in ecological and human (biological, socioeconomic, and sociocultural) interactions is highlighted by the Coupled Human and Natural Systems model (Kassam 2010a, b; Lassoie and Sherman 2010; Walsh and McGinnis 2008; Liu et al. 2007a, b, c; Buntaine et al. 2007; McPeak et al. 2006; Olsson et al. 2004; Machlis et al. 2005; Berkes et al. 2003;

Folke et al. 2002; Liu et al. 2001; Machlis et al. 1997). It is also referred in the literature as coupled social (or human) and ecological systems, which will be used for the
remainder of the book.

A good example of this coupled systems interaction is a comprehensive study
by Liu et al. (2001) that illustrated how a government policy to help protect panda
habitat in the Wolong Panda Reserve in Sichuan, China by paying local people to
monitor illegal cutting, was actually contributing to more illegal cutting and a
higher rate of panda habitat destruction. This was due to a policy that rewarded
individual households to guard against illegal cutting. As a result, households
began splitting up to form new households to take advantage of the subsidy. More
households required more wood, which led to more cutting, thus illustrating an
unintended outcome arising from a policy decision involving a coupled social and
ecological system.

This body of coupled systems research along with the work of many others (e.g.,
ADB 2008; Brushett and Osika 2005; Salick et al. 2005; Flower 2004; Buffetrille
2003; Cook 1991; Price 1989; Cruikshank 1985) points out the importance of a
comprehensive approach to evaluating policy, development, and management practices. This should include analysis of environmental, socioeconomic, and sociocultural spheres, and how they interact and influence each other (Liu et al. 2007a).
Hence when policy and research professionals study road development and its
impacts, taking a coupled systems approach will help provide a more comprehensive understanding of how roads impact and influence humans and the environment.
The case studies from empirical research (Beazley 2013) included in this book
(Chap. 3) are intended to highlight this coupled social and ecological system of
expanding road networks in Nepal.

1.3   A Brief Overview of Road Impacts

Roads can have both positive and negative influences that have far ranging, unanticipated, and unintended consequences on environmental, socioeconomic, and
sociocultural spheres (Table 1.2). Many of these consequences are a result of
focusing on only one of the spheres in isolation, when in fact they are all interrelated. The following is a brief summary of road impacts. These impacts will be
explored in much more detail in Part III: The Challenges and Impacts of Road
Building in the Himalayas.

Environmental impacts of roads are mostly negative, in that they contribute to
habitat fragmentation, soil erosion, air, noise, and water pollution, the introduction
of invasive species, and the death of wildlife, livestock, and humans (Forman et al.
2002). However, a few positive environmental impacts of roads have been reported,
such as road verges providing habitat for birds, butterflies, and other small animals
(Munguira and Thomas 1992; Adams 1984; Oxley et al. 1974), and roads in forested areas enhancing plant growth by increasing light availability and acting as fire
breaks (Brown 2001).

Table 1.2 Positive and negative impact of road construction in the environmental, socioeconomic, and sociocultural spheres (Beazley 2013)

Road impact effects	Positive	Negative
Environmental	Road verges provide habitat for birds, butterflies, and other small animals	Habitat fragmentation
	Roads in forested areas can enhance growth by increasing available light and act of fire breaks	More pollution: air, water, noise
		Alteration of surface runoff
		Alteration of ground water recharge
		Mass wasting: erosion, landslides
		Road kill
		Invasive species
		Urban heat islands
Socio-economic	Increased mobility	Change in land values that usually takes advantage of the poor
	Better access to: markets, employment opportunities, schools, and health care	Relocation or loss of land
		Increased disparity between rich and poor
		Increased in-migration that takes jobs from locals
Socio-cultural	Increased mobility	Social conflict due to in-migration
	Better access to: family and friends, sacred spaces, cultural sites, and spiritual teachers	Increase in AIDS and sexually transmitted diseases
		Increase in human trafficking and commercial sex
		Increase in crime
		Loss of traditional culture
		Loss of sacred space

In theory, roads lead to socioeconomic development by increasing connectivity, which leads to cheaper transportation costs, and shorter travel times. In the case of rural mountain farmers, these benefits allow them to take advantage of better access to markets (Jacoby 2000; Richards 1984) and expand their production. In addition, by enhancing connectivity roads give people the opportunity to travel outside their local area to seek employment, access education and health care facilities, and enhance their social capital, which in turn can have positive impacts on poverty reduction.

Relatively few studies have examined the sociocultural impacts of road development and in particular, how such impacts are interrelated with both socioeconomic and environmental impacts. Four case studies that addressed sociocultural impacts highlight the negative effects of roads on indigenous groups: one along the Alaska Highway (Cruikshank 1985) and the others in South America (World Bank 2006; Ayers et al. 1991; Gilio 1986). Another sociocultural impact that is well documented is the increase in the sex trade, HIV, and sexually transmitted infections associated with roads, increased mobility, and migrants. These are often called 'highway diseases' due to the well-established connection between the commercial sex trade and roads (ADB 2008; Brushett and Osika 2005; UNDP 2006).

1.4   Global Road Building

"Few forces have been more influential in modifying the earth than transportation." This quote by Edward Ullman (1956: 862) one of the leading transportation geographers of the twentieth century is a statement that most of us, on reflection, would probably agree with but are rarely aware of in our day-to-day activities. It is virtually impossible to calculate the total number of roads on the face of the earth and estimates from various sources vary greatly. The World Bank statistics for total road networks by country[1] is missing data on different countries for given years, but for the country data available for 2011 (82 out of 216 countries) the total is 28,840,974.05 km. Obviously if all the data were available this number would be much higher. The total length of roads in these 82 countries alone is enough to encircle the earth 720 times,[2] or reach the moon and back to earth 37.5 times.[3] The CIA for 2013 lists the total world road network (227 countries) at 64,285,009 km (CIA n.d.).

The Socioeconomic Data and Applications Center in NASA's Earth Observing System and Information System has developed the Global Roads Open Access Data Set, version 1 (1980–2010) (Ubukawa et al. 2014).[4] According to their calculations there is a total of 9,100,000 km of roads in 221 countries. The world map of roads (Fig. 1.1) produced from their data reveals that apart from northern Canada, Greenland, and the arctic, roads cover much of the earth's surface.

Global road building is having a complex impact on the planet, and the physical manifestation of roads on the earth's surface has not only huge environmental impacts but socioeconomic and sociocultural impacts as well. When one considers the scale of impacts in all three spheres that travel beyond the areas immediately adjacent to the road surface, global road building must be acknowledged as one of the most profound and influential human activities on earth. Renowned landscape architect John Brinckerhoff Jackson (1980: 122) summed up his study of roads by saying that they are "now the most powerful force for the destruction or creation of landscapes that we have." He continues, "…the really significant thing about the road was how it affected the landscape; how it started out as a wavering line between fields and houses and hills and then took over more and more land, influenced and changed a wider and wider environment until the map of the United States seemed nothing but a web of roads and railroads and highways" (Jackson 1980: 122).

1.5   A Brief History of Road Development in Nepal

On March 21, 1770, King Prithvi Narayan Shah declared Kathmandu the capital of Nepal. From that time until 1927, the only way in and out of the Kathmandu Valley was on trails by foot or animal. Transport of goods was done either by human

---

[1] http://search.worldbank.org/data?qterm=roads&language=EN&op=.

[2] The circumference of the earth is 40,075 km.

[3] The distance to the moon is 384,400 km.

[4] http://sedac.ciesin.columbia.edu/data/collection/groads/sets/browse.

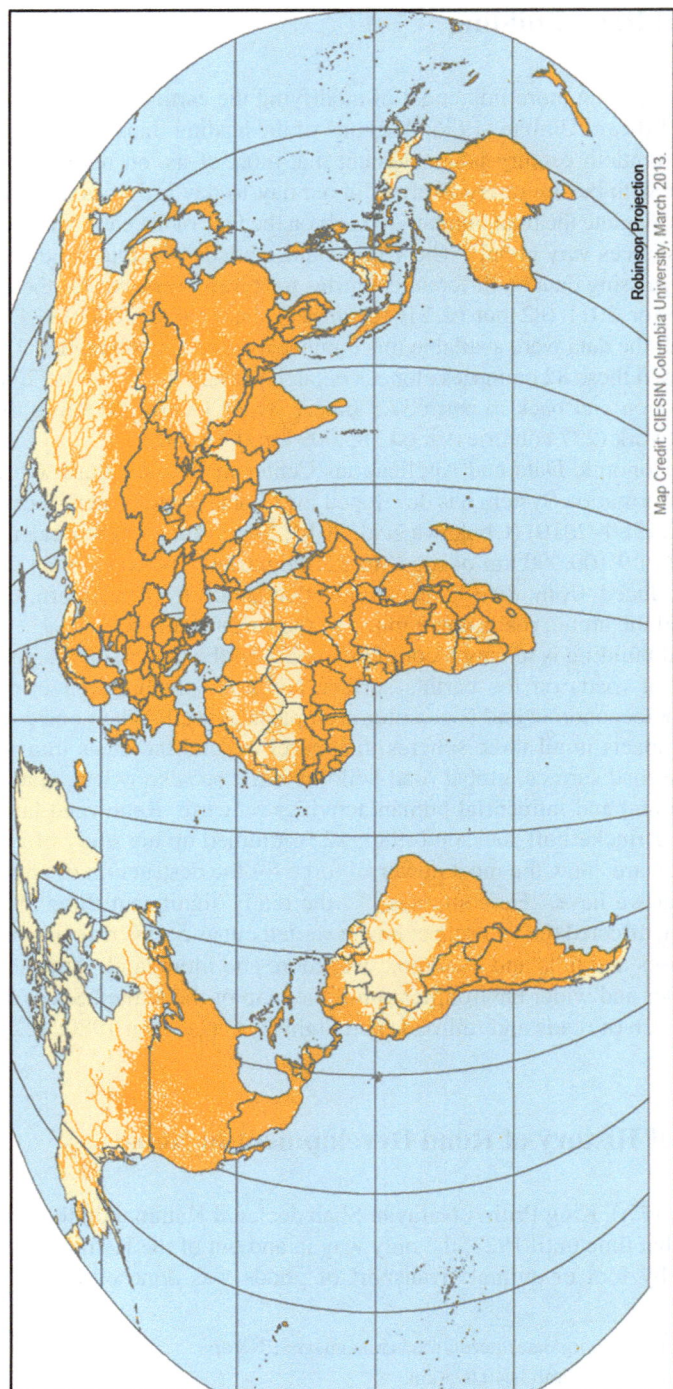

Fig. 1.1 Global roads (The Global Roads Open Access Data Set, Version 1 (gROADSv1)) (http://sedac.ciesin.columbia.edu/data/set/groads-global-roads-open-access-v1/maps)

porters or on the backs of mules or ponies. Starting in 1924 a plan was initiated to connect Kathmandu to India on its southern border by three interconnected transportation systems: a ropeway[5] from Kathmandu to Dhursing, a roadway from Dhursing to Amlekhgunj, and railway from Amlekhgunj to Raxaul on the Nepal-India border.

The first all-weather gravel road outside the Kathmandu Valley was constructed in 1929 and was only 42 km long (Shrestha 1980). There seems to be some debate over when the first automobile arrived in Kathmandu, but until the 1950's, they were all carried over the hills and into the Kathmandu Valley on the backs of porters (Fig. 1.2). "Cars brought mainly by the Rana or Shah nobility were brought to Calcutta by ship, driven up to Bhimphedi and then carried over the mountains by porters" (Bhujel 2014). This often required negotiating stream crossings (Fig. 1.3) and other obstacles along treacherous narrow trails.

One of the few surviving car porters of the 1930s, Dhan Bahadur Gole, carried his first car, a Daimler, at the age of 17 with 64 other porters. In total he carried 30 cars over a 20-year period. His main preparation for the car carrying trips was to weave two pair of straw sandals.

Fig. 1.2 Porters carrying a car over the last ridge into Kathmandu (http://visitnepal-yogesh.blogspot.com/2012/02/100-years-ago-in-nepal.html

---

[5]A ropeway uses rope or cable suspended between towers to carry loads over steep terrain and across rivers. The first rope way was installed in Nepal in the 1920s. http://library.uniteddiversity.coop/Ecological_Building/aerial_ropeway.pdf.

Fig. 1.3 Porters carrying a car across a river heading to Kathmandu (Volkmar Wentzel http://ngm.nationalgeographic.com/ngm/exploration/postcard06.html)

"A pair was never enough, sometimes we would wear down two pairs of slippers even before we reached Chitlang, and so we had to make more on the way," …"We didn't even know the model of the cars we were carrying, we just called them 32, 64, 96 depending on the number of people carrying them," Dhan Bahadur recalls. (Bhujel 2014)

While no one agrees as the to exact year, it appears that the first car arrived in Nepal in the early 1900s. According to Thapa (2013: 4):

First motorized vehicle entered Kathmandu Valley in year 1901 on shoulders of men for the service of ruling Elites of that time. The time of entering first vehicle in Kathmandu differs with the previous Status papers, but this is the correct one. Latest photographic evidences of vehicle carrying by men are the vehicle gifted by Adolf Hitler to king Tribhuvan of Nepal in 1942. But the vehicles were running on Kathmandu roads well before. Number and types of vehicles increased in Nepal after the opening of Tribhuvan Rajpath (TRP), constructed by Indian Assistance in 1956, linking Terai and Indian border with Kathmandu valley. [sic]

However, another source claims it was 1908 (Vehicles Grew 15 Times in 20 Yrs 2011) and yet another cites 1916 (Bajracharya et al. 2006). Regardless of which date is correct the early 1900s coincides with the era of the "First Car" in Nepal as noted in country's National Museum.

Indeed, the story of the First Car is quaint if not quirky, like most of Nepal's history. According to Bharat Raj Rawat, Museologist of the National Museum, when the Rana ruler saw his first car, he was smitten by road fever and a quest for modernity. Leveraging his good relations with the British government then ruling over colonial India, he arranged to have a Hudson motor car transported by freighter all the way from Australia to Bangladesh. From there it was put on a railroad flatcar. When the tracks ran out in Delhi, porters removed the car's wheels, hoisted the chassis onto two long wooden poles, and humped it over the

Himalayan foothills to the royal palace in Kathmandu. The process of building a road and shipping a car to the middle of nowhere went smoothly enough, but the car arrived without "petrol or spare tyres." While waiting for these annoyances to be shipped, a coterie of royal pushers put their backs into the car's rear end, with the King and Queen sitting in the front, proud as pashas, their armed bodyguards keeping a watchful eye from the backseat. In this way, the First Car was paraded up and down the New Road to the delight of cheering throngs of children, jumping up and down behind in the dust. (Pohl n.d.)

Nonetheless, before 1950 motorable roads in Nepal were virtually non-existent (Paudyal 1998). In 1949, air links were established between Kathmandu and India but it was not until 1956 that Kathmandu had a road usable by trucks that linked to the existing road in the flat Terai area connecting Kathmandu to the railhead and India. Consequently, for almost 30 years, from 1927 until the completion of this section of road, the only way to transport goods into or out of the Kathmandu Valley was by the ropeway, human porter, or animal (Shrestha 1980). This is highlighted by the fact that until 1956 there was only approximately 600 km of roads in and around the few main cities of Nepal, and more than half were "fair weather only" roads (Shrestha 1981 cited in Paudyal 1998). In 1961, a section of road was completed from Amlekhgunj to Raxaul making the road link from India to Kathmandu complete (Shrestha 1980).

The cars that did make it into Nepal before roads were intended for royalty and the upper class elites who used them in a very limited area of Kathmandu (Shrestha 1980). The presence of a Ford dealer in the Lazimpath area of Kathmandu in the 1930's (Fig. 1.4), speaks to the wealth of the royal family and their upper class friends.

Additional interesting vintage vehicle photos and a video of a car being carried by porters in Nepal can be found at the History of 'Nepal Transport Service' Facebook website.[6]

While many different factors were involved in the overall plan to connect the capital of Nepal in the Kathmandu Valley to the Indian border, a distance of only 115 km, one factor that cannot be ignored is the difficulty and expense of building roads in mountainous areas. Both the road and rail sections of the links in the flat Terai area were complete by 1927, but it was another 30 years before there was a road connecting those links through the mountainous terrain and over the pass into the Kathmandu Valley. This was the first road of national importance completed in Nepal as it directly connected the two major economic centers, the Kathmandu Valley and Raxaul, and adjoining areas on the India border (Shrestha 1980). The same year the link was completed with India in the south plans were being made to connect Kathmandu with China in the north. In 1961, King Mahendra signed an agreement with the Chinese government in Peking (Beijing) to construct a road from Kathmandu to the Chinese border in Tibet at Kodari. The 114 km road was begun in 1961 and completed in 1966 and was quite controversial at the time because of the political implications of Kathmandu being connected by road to both India and China (Raj 1978). This road was subsequently named the Arniko Highway (a.k.a. Araniko Highway and Arniko Rajmarg), after the famous Neplai artist, sculptor, and architect Arniko.[7]

---

[6] https://www.facebook.com/pages/History-of-Nepal-Transport-Service/149043701774885.

[7] At the age of 16 Arniko was sent by one of the Kathmandu Kings, Jaya Bhim Dev Malla on the request of Mongolian Emperor Kublai Khan's spiritual teacher Phags-pa to build a golden stupa in Lhasa, Tibet. With a group of 80 artisans he completed his work so well that the Emperor requested

Cars for the upper class: the authorized Ford dealer in Lazimpath, probably in the 1930s. Cars were carried over the mountain trails on bamboo cross-poles by teams of 64 porters.

Fig. 1.4 The caption reads, "Cars for the upper class the authorized Ford dealer in Lazimpath probably in the 1930s. Cars were carried over the mountain trails on bamboo cross-poles by teams of 64 porters" (http://meropost.com/view/post:6727)

Around the same time, to avoid having to cross into India, plans were made for an east–west transportation road network within Nepal. India, not wanting to lose its advantage from the existing road system, showed no interest in helping build the road until China began building a portion of it, at which point India agreed to take over construction of the remaining sections (Blaikie et al. 1976). It has been suggested that all three of these roads, the first major roads in Nepal, received aid from India and China for strategic reasons (Raj 1978; Blaikie et al. 1976) rather than purely as economic development aid. This strategic motivation behind road building is common and has been observed by other authors such as Kreutzman (2000) in

him to come again for other works. In total he is credited with building two Confucian shrines, a Taoist palace, the Archway of Yungtang, nine Buddhist monasteries, and the White Pagoda of Miaoying Temple in Peking (Beijing). The latter was so highly thought of that it was honored as a national historical treasure. He is credited with bringing the pagoda style of Kathmandu architecture to China. His talents also included sculpture and painting; he was entrusted with painting a portfolio of Chinese Emperors, which were held in equally high regard. He was eventually made a minister in the Emperor's court and bestowed with the title of Liang Guo Gong (Duke), which included an estate of 15000 acres, 1000 serfs, and 100 head of livestock near Peking. He died in China in 1306 at age sixty-two. He was further honored by being one of the few foreigners whose biography became part of the history books of the Imperial Library (Shrestha 2011).

other mountainous areas of Asia during the same period. The importance of road building from the mid-1960s through the mid-1970s is evident from the amount spent on road construction, which was greater than the total for all other government projects (Blaikie et al. 1976). Over a 20-year period from 1956 when the First Five Year Plan was initiated to 1976 during the Fifth Five Year Plan, road length had expanded from 625 km (0.40 km per 100 km^2) to 4136 km (<2.20 km per km^2) (GoN/MPPW/DoR 2008; HMGN/MWT/DOR 1985) and effectively produced a north-south and east-west road network.

Nonetheless, accessibility for the average rural Nepali was still very limited especially in the hills and mountainous areas. The east-west road corridor was restricted to the flat Terai area in the south and the north-south corridor only ran through the Kathmandu area in the east. During the 1980–1990s, the government targeted minimum transport facilities to rural areas as a priority giving grants to local governing bodies at the district and village level as one strategy to help attain this goal. However, by the end of the 1990s, at least 20 of Nepal's 75 districts still lacked connection by vehicle roads (Paudyal 1998). By 2007, there was almost 19,000 km of roads (12.79 km per 100 km^2) but only 9399 km of that total was considered part of the Strategic Road Network (SRN), the remainder being local roads. Of the total SRN roads 45% were blacktop, 33% earthen, and 22% gravel (GoN/MPPW/DoR 2008).

More recent comprehensive information about Nepal's Road Network can be found in the Department of Roads' 2013 Status Paper on Road Safety in Nepal (Thapa 2013). The National Road Network (NRN), the sum total of all the roads in Nepal (2013), equals 62,579 km of which 11.4% are bituminous, 26.3% are gravel, and 62.3% are earthen roads. Under the NRN there are two broad main categories, the Strategic Road Network (SRN), which has a total of 11,636 km and the Local Road Network (LRN) with a total of 50,943 km (Table 1.3).

The SRN, "the backbone of the National Road Network", is made up of 21 National Highways and 208 Feeder Roads, which "are the main national arteries providing inter-regional connections and links to regional and district headquarters, international borders, key economic centers, touristic centers and the major urban roads" (Thapa 2013: 2). The design, construction, and maintenance of the SRN are the responsibility of the Department of Roads (DoR) under the Ministry of Physical Infrastructure, Works, and Transportation Management (MoPIWTM) also known as the Ministry of Physical Infrastructure and Transport (MoPIT) (Figs. 1.5 and 1.6).

Table 1.3 Total road network of Nepal, 2013 (Adapted from Thapa 2013)

Type of road	Bituminous	Graveled	Earthen	Total	Population per km (26,620,809)	km per km (147,181)
SRN[a]	5574	1888	4173	11,636	2288	7.90
Local[b]	1575	14,601	34,766	50,943	522	34.61
Total	7149 (11.4%)	16,489 (26.3%)	38,939 (62.3%)	62,579 (100%)	425	42.51

[a]Summary of Rural Roads, DOLIDAR, 2013
[b]Road Statistics, Department of Roads, 2010

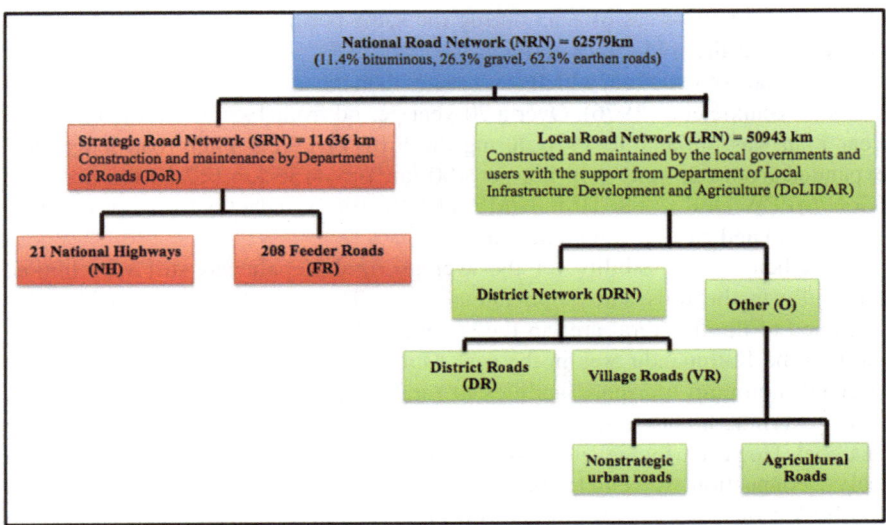

Fig. 1.5 National Road Network Diagram 2013 (Adapted from Thapa 2013)

The LRN consists of District Roads (DR), Village Roads (VR), Agricultural Roads (AR), and Nonstrategic Urban Roads (NUR). Design, construction, and maintenance of the LRN are the responsibilities of local governments, including the District Development Committee (DDC), the Village Development Committee (VDC), and Municipalities, under the direction of the Department of Local Infrastructure Development and Agriculture Roads (DoLIDAR) (Fig. 1.7) of the Ministry of Federal Affairs and Local Development (MoFAaLD).

Construction and maintenance of the LRN often involves the participation and employment of local users and villagers (Thapa 2013) (see Fig. 1.5). The recent building of rural roads under DoLIDAR was planned and implemented according to the 2005 Integrated Rural Accessibility Planning Guidelines (HMGN/MLD/DoLIDAR 2005). Under the DoLIDAR plan, local Village Development Committees (VDC) work with the District Development Committees (DDC) to prepare a District Periodic Plan that covers the development needs of the district over a least a 5-year period including a District Transport Master Plan (DTMP). This process of decentralization to the village level has been an on-going process in Nepal for some time with the recognition that the local inhabitants benefit more from projects that they identify and participate in rather than those planned from the top down. The 2001 National Transport Policy states this concept of decentralized governance and local development of transport as one of its three main strategies (HMGN/MPPW 2001). Consequently, the new Integrated Rural Accessibility Planning Guideline (2005) was instituted to further this process by giving local governments even more involvement in infrastructure projects, which stated (HMGN/MLD/DoLIDAR 2005: 16):

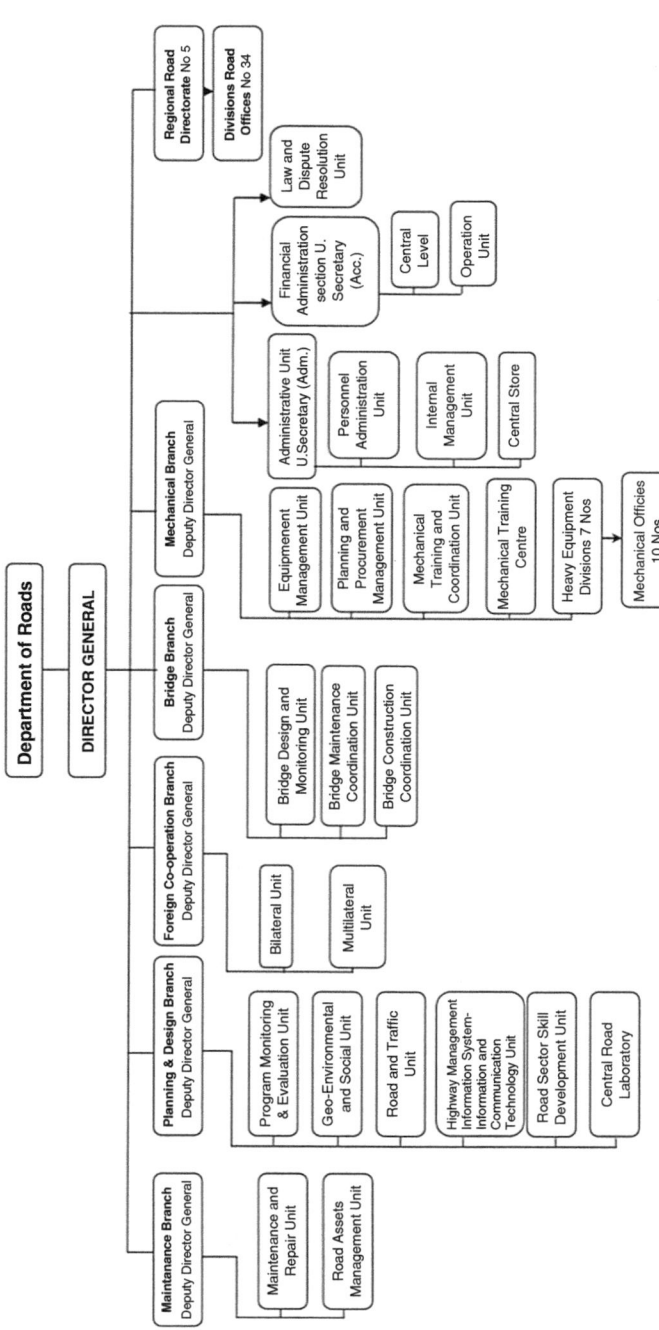

Note: This is a Present Functional Organization of Department of Roads

Fig. 1.6 Organizational chart Department of Roads (GoN/MoPIT/DoR 2016: 254)

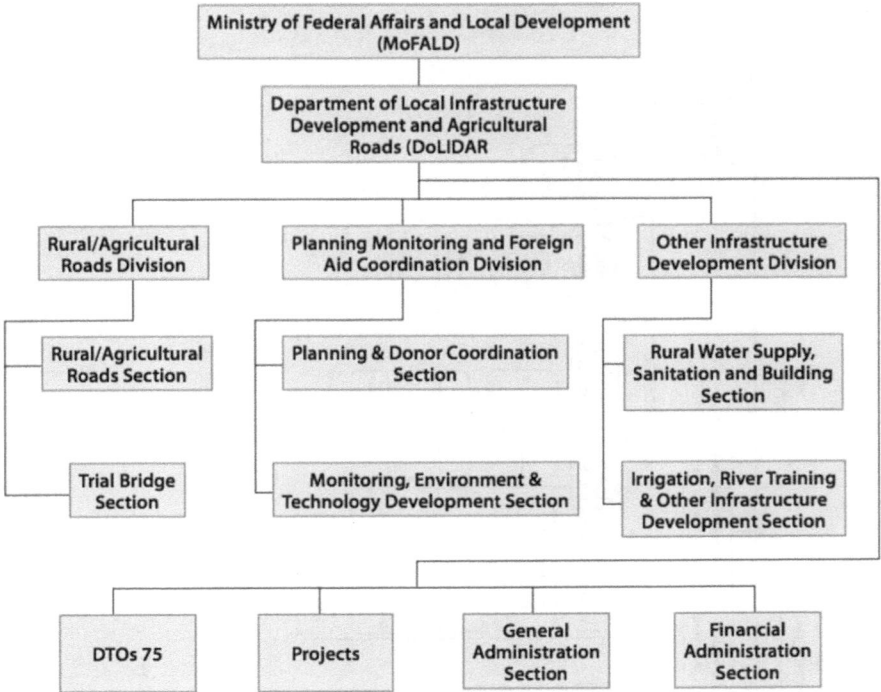

Fig. 1.7 Organizational chart Department of Local Infrastructure Development and Agriculture Roads (GoN/ MoFAaLD /DoLIDAR 2016: 5)

This policy covers infrastructure for local transportation, irrigation and river control, small hydro power and alternate energy, drinking water, sewerage and sanitation, housing, building and urban development, management of solid waste and social infrastructure including government offices, health, education etc. The policy aims to increase the participation of local people in physical and social infrastructure development to enhance the social services, economic opportunities and mobilization of local resources

Under this strategy, the goal is to provide access in rural areas to all-weather motorable roads within 4 h walking distance in the hills and 2 h in the Terai. It was estimated in 2007 that 39% of the population in the hills do not have access within 4 h and 13% in the Terai do not have access within 2 h (Shah 2007). In 2007, there were still 12 district headquarters, of a total 75 that did not have a road connection (GoN/MPPW/DoR 2008). However, by 2009, that number had been cut in half (Sitaula 2009) and by 2013 only two district headquarters remained unconnected (Thapa 2013). Funds for road construction come from several sources including the Central Government, the Roads Board, and donor agencies. Donor agencies such as the Asian Development Bank (ADB) and the World Bank (WB) contribute through multi-lateral loans and bi-lateral grants making up the majority of the contributions at 70% of the total (Pande 2006).

1.6   Local Road Construction

As acknowledged previously there are many institutions that oversee and engage in road construction in Nepal. The Local Road Network is developed from the District Transport Master Plan, which is written with the participation of the Village Development Committees. Money is allocated to the VDCs for road construction projects, which are then carried out at the local level according to the local politics of the area. While this empowerment of local governments may better address the needs of the local people it can also lead to corruption when local politicians who are often also the local contractor influence road alignments. This can lead to *ad hoc* road construction and environmentally detrimental construction techniques (Dahal et al. 2010).

1.7   Green Roads

The *Green Roads Concept* (GRC) promotes local participation under technical guidance in village road construction projects using environmentally friendly construction techniques (Mulmi 2009; Klatzel 2000). Many village Green Roads have been built in Nepal with the support of INGOs such as the Swiss Agency for Development and Cooperation (SDC) and German Technical Cooperation (GTZ). The GRC strives to reduce poverty by employing local villagers and to practice road construction techniques that avoid the detrimental environmental impacts associated with mechanized road building techniques, such as improper cut and fill practices. This is often referred to in the literature as a labor based, environmentally friendly, and participatory (LEP) technique (Mulmi 2009).

1.8   The Importance of Roads in Nepal

Nepal is a landlocked country bordered by India on the west, south, and east and by China on the north. Kathmandu has an international airport and there are numerous small airports and landing strips throughout the country. There is one 52 km operating railway system but due to its antiquated and poorly maintained condition it is not considered a substantial transport vector (Thapa 2013). The majority of Nepal's transportation of goods and people has traditionally relied on its system of roads and trails and their continuing expansion (Table 1.4 and Fig. 1.8).

Vital supplies such as petroleum typically come by road from India via the southern road system to Kathmandu. Chinese goods arrive via the one paved northern road, the Arniko (a.k.a. Araniko) Highway. A recently completed (2012) second road link north through Rasuwa District will provide a second northern link from Kathmandu to China.

Table 1.4 Comparative influenced population and density (1998–2015/16) (Adapted from GoN/MoPIT/DOR 2016: 1)

Year AD	Description	Length			Total	Influenced population (no. per km)	Density km/100 sq. km
		BT	GR	ER			
1998	9th (5 year plan) 2054	2905.00	1656.00	179.00	4740.00	3901.08[a]	3.22
2000		2974.00	1649.00	171.00	4794.00	3857.13[a]	3.26
2002	10th (5 year plan) 2058	3028.74	1663.84	168.68	4860.96	4762.73[b]	3.30
2004		3494.73	883.51	614.49	4992.73	4636.23[b]	3.39
2006/07		4258.20	2061.7	3079.48	9399.38	2463.08[b]	6.39
2009/10		4952.11	2065.15	3817.76	10835.02	2136.72[b]	7.36
2011/12		5573.55	1888.49	4173.55	11635.58	2287.88[c]	7.91
2013/14		6368.98	1735.49	4389.47	12493.94	2130.70[c]	8.49
2015/16		6823.43	2044.22	4030.55	12898.20	2063.92[c]	8.76

[a]Population Census 1991
[b]Population Census 2001
[c]Population Census 2011

While significant progress has been made (see Table 1.4) road density still remains low compared to other south Asian countries.

> Among the South Asian Countries, Nepal has a very low road density, not only in terms of serving the population but also in providing accessibility to various parts of the country. Although the strategic roads constitute about 20 percent of the National Road Network, it plays a very important role in terms of the movement of people and freight. The strategic roads have traffic volume in comparison to district roads. (Thapa 2013: 2)

The figures can be misleading in terms of the operability of the roads as both SRN and the LRN suffer from frequent closures of sections of road due to landslides. The LRN in particular is susceptible to long periods of inoperability as a result of poor planning, construction, and maintenance.

Concerning the LRN Thapa (Thapa 2013: 2–3) comments:

> Actually, there was no inventory of these roads. DOLIDAR has recently conducted an inventory survey and found that there is some 51,000 km of Local Road Network in the country as of 2013 (Summary of Rural Roads, DOLIDAR, 2013). These roads were constructed to open up access to remote/rural areas as quickly as possible but without giving due consideration to the operability and sustainability of roads thus constructed. The geometry of these roads is poor and some serious environmental problems are created by these roads. About only 40 percent of the network is serviceable. However, the Department of Roads (DoR) also looks after approximately 10,000 km of main roads of this local network (Road Statistics, Department of Roads, 2010), which are mostly serviceable.
>
> With the combined lengths of SRN and local road networks of 62,579 km, the road density km per 100 sq. km is 42.51 and influenced population number per km is 425. If the serviceability is considered, the road density drops down drastically.

The recently published Statistics of Strategic Road Network (SSRN) (GoN/MoPIT/DoR 2016: 8) states that the total SRN is now 12898.20 km. Table 1.5 shows the road lengths by region (BT=Bituminous, GR=Graveled, ER=Earthen, UC=Under Construction, PL=Planned). The SRN is composed of 53% blacktop, 31% earthen,

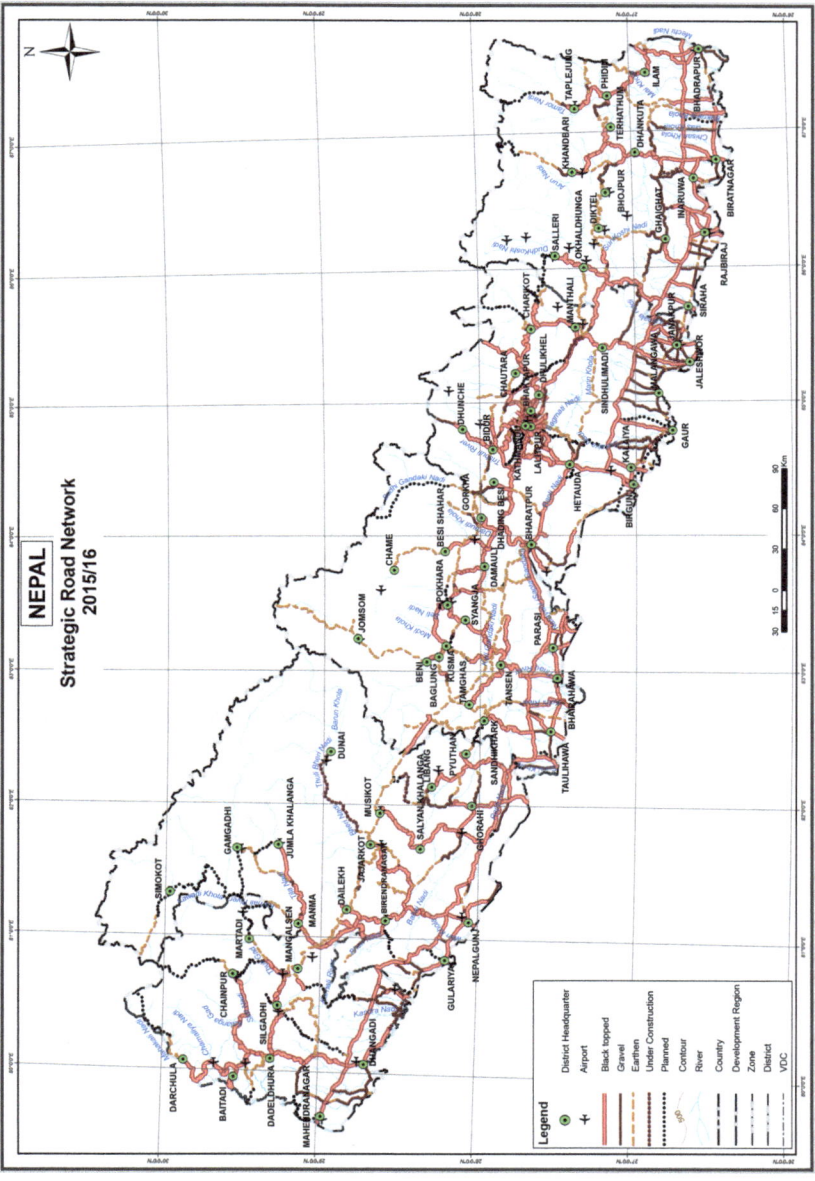

Fig. 1.8 Strategic Road Network 2015/2016 (GoN/MoPIT/DoR 2016: 11)

Table 1.5 SRN lengths by geographic region with road category (in km)

	Region	BT	GR	ER	Total	UC	PL
National Highway	Eastern	634.15	68.58	115.00	817.73	29.00	80.00
	Central	847.37	28.81	0.00	876.18	0.00	142.80
	Western	478.16	0.00	0.00	478.16	0.00	0.00
	Mid-Western	724.44	8.00	2.66	735.10	0.00	0.00
	Far-Western	518.03	0.00	50.00	568.03	0.00	93.30
	Sub-total	**3202.15**	**105.39**	**167.66**	**3475.20**	**29.00**	**316.10**
Feeder Road	Eastern	633.67	311.50	394.06	1339.23	0.00	131.50
	Central	1115.71	548.98	478.54	2143.22	73.90	375.09
	Western	726.62	98.06	1213.49	2038.17	8.00	195.01
	Mid-Western	563.55	238.67	500.00	1302.22	156.40	366.00
	Far-Western	281.03	149.12	332.30	762.45	0.00	434.00
	Sub-total	**3320.58**	**1346.33**	**2918.39**	**7585.29**	**238.30**	**1501.60**
Mid-Hill Road	Eastern	48.00	128.00	248.00	424.00	0.00	0.00
	Central	47.00	47.00	54.00	148.00	33.00	80.00
	Western	62.00	13.00	140.00	215.00	0.00	27.00
	Mid-Western	0.00	32.00	296.00	328.00	0.00	0.00
	Far-Western	20.00	0.00	61.00	81.00	0.00	0.00
	Sub-total	**177.00**	**220.00**	**799.00**	**1196.00**	**33.00**	**107.00**
Postal Road	Eastern	23.00	63.00	50.00	136.00	16.00	9.00
	Central	39.00	147.00	63.00	249.00	16.50	12.50
	Western	0.00	58.50	2.00	60.50	0.00	8.00
	Mid-Western	24.00	52.00	24.00	100.00	3.00	8.00
	Far-Western	37.70	52.00	6.50	96.20	27.50	3.00
	Sub-total	**123.70**	**372.50**	**145.50**	**641.70**	**63.00**	**40.50**
	Grand total	**6823.43**	**2044.22**	**4030.55**	**12898.20**	**363.30**	**1965.20**

16% gravel (Fig. 1.9). A long time goal of the DoR Master Plan has been to connect all 75-district headquarters by road. There only remain two that have not yet been connected (Table 1.6). Dolpo District only has 56 km of SRN and Humla has 180 km (GoN/MoPIT/DoR 2016: 8). Both districts are in the western part of Nepal and share northern borders with Tibet. They have very mountainous areas where road building is extremely difficult. In general northwestern Nepal's development has been slower than other areas due to lack of priority (until recently) and its relative remoteness and inaccessibility (Fig. 1.10). The Statistics of Local Road Network (SLRN) 2016 show a total of 57632 km with 3.5% black top, 22.2% gravel, and 74.3% earthen (Fig. 1.11) (GoN/MoPIT/DoLIDAR 2016: 1).

Since the end of the People's War and the signing of the Peace Accord in 2006, Nepal has been undergoing a huge change in its government including the mandate to write a new constitution by May 2010. The deadline was extended several times until finally in May 2012 Prime Minister Bhattari dissolved Parliament for failure to meet the latest deadline. In November 2013 elections confirmed a new constituent assembly to renew the task of writing a new constitution. Sushil Koirala (2014) was elected in February 2014 with a mandate to finish the new constitution within 1 year.

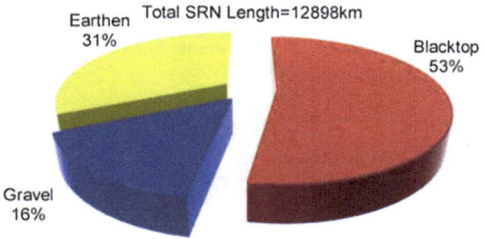

Fig. 1.9 Percentage of SRN according to surface type (GoN/MoPIT/DoR 2016: 1)

Table 1.6 Remaining districts headquarters not connected by road (GoN/MoPIT/DOR 2016: 6)

	District headquarters not connected with road	
SN.	District head-quarters	District
1	Dunai	Dolpa
2	Simikot	Humla

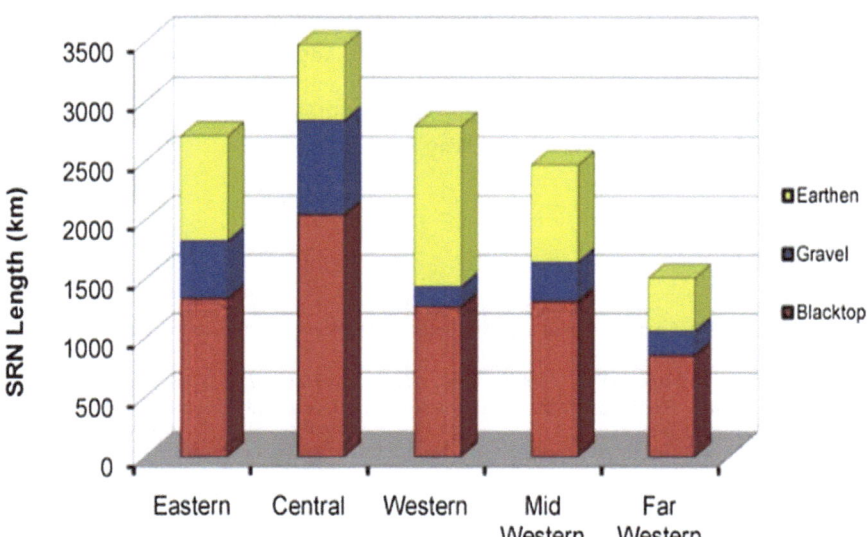

Fig. 1.10 SRN road lengths in different geographic regions of Nepal (GoN/MoPIT/DoR 2016: 1)

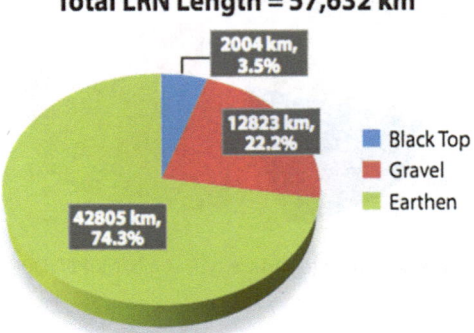

Total LRN Length = 57,632 km

Fig. 1.11 Percentage of different road surfaces of the Local Road Network (LRN)

The deadline passed with no new constitution. Two months later, in April 2015, Nepal experienced a devastating earthquake with numerous serious aftershocks. Amid the chaos of the aftermath of the earthquakes a new constitution was promulgated on September 20, 2015 replacing the interim constitution of 2007. This was followed by unrest and a series of boycotts and protests by citizens unhappy with the conditions of the new constitution and a 4 month blockade along Nepal's southern border cutting off Nepal's only supply of petroleum (from India). In October 2015 a new Prime Minister, Khadga Prasad Oli, replaced Koirala. Oli served until August 2016 when former Prime Minister Pushpa Kamal Dahal replaced him. Dahal had previously been Prime Minister from August 2008 to May 2009. The details of the frequent government changes and the affects of both the earthquakes and the protests and blockade following the promulgation of the new constitution are explained in detail in Part IV, The Way Forward.

References

Adams, L. W. (1984). Small mammal use of an interstate highway median strip. *Journal of Applied Ecology, 21*, 175–178.

Asian Development Bank (ADB). (2008). *ADB, roads, and HIV/AIDS: A resource book for the transport sector*. Retrieved from http://www.adb.org/Documents/Books/ADB-HIV-Toolkit/ADB-HIV-Toolkit.pdf.

Ayers, J. M., De Magalhes Lima, D., De Souza Martins, E., & Barrieros, J. L. K. (1991). On the track of the road: Changes in subsistence hunting in a Brazilian Amazonian village. In J. G. Robinson & K. H. Redford (Eds.), *Neotropical wildlife use and conservation* (pp. 82–92). Chicago: University of Chicago Press.

Bajracharya, D., Bhuju, D. R., & Pokhrel, J. R. (2006). *Science, research, and technology in Nepal*. United Nations Educational, Scientific and Cultural Organization (UNESCO) Report, Kathmandu, Nepal: UNESCO. Retrieved from http://unesdoc.unesco.org/images/0014/001461/146117e.pdf.

Beazley, R. E. (2013). *Impacts of expanding rural road networks on communities in the Annapurna Conservation Area, Nepal*. M.S. Thesis, Department of Natural Resources, Cornell University, Ithaca, NY.

Berkes, F., Colding, J., & Folke, C. (2003). *Navigating social-ecological systems: Building resilience for complexity and change*. Cambridge, UK: Cambridge University Press.

Bhujel, S. K. (2014, April 21–27). Nepal's last car carrier. *Nepali Times*. Retrieved from http://nepalitimes.com/article/Nepali-Times-Buzz/Nepal-last-car-carrier,1216.

Blaikie, P., Cameron, J., Feldman, D., Fournier, A., & Seddon, D. (1976) *The effects of roads in West Central Nepal*. A report to the Economic and Social Committee for Overseas Research, Ministry of Overseas Development. East Angila, UK: Overseas Research Group, University of East Anglia.

Brown, D. (2001). Characterizing the human imprint on landscapes for ecological assessment. In M. E. Jensen & P. S. Bourgeron (Eds.), *A guide for integrated ecological assessments* (pp. 404–415). New York: Springer-Verlag.

Brushett, S., & Osika, J. S. (2005). *Lessons learned to date from HIV/AIDS transport corridor projects*. World Bank Global HIV/AIDS Program. Retrieved from http://siteresources.world-bank.org/INTHIVAIDS/Resources/375798-1103037153392/Transport.pdf.

Buffetrille, K. (2003). *The evolution of a Tibetan Pilgrimage: The Pilgrimage to A myes rMa chen Mountain in the 21st Century*. 21st Century Tibet Issue. Symposium on Contemporary Tibetan Studies. Taipeh, Taiwan. Retrieved from www.case.edu/affil/**tibet/tibetan**Nomads/documents/Taiwan_art.doc.

Buntaine, M. T., Mullen, R. B., & Lassoie, J. P. (2007). Human use and conservation planning in alpine areas of northwestern Yunnan, China. *Environment, Development, and Sustainability, 9*(3), 305–324.

Central Intelligence Agency (CIA). (n.d.). *The World Fact Book*. Retrieved from https://www.cia.gov/library/publications/resources/the-world-factbook/fields/2085.html.

Cook, C. (1991). Social analysis in rural road projects. In M. K. Cerena (Ed.), *Putting people first: Sociological variables in rural development (chapter 11)* (pp. 397–494). Cambridge, UK: Oxford University Press.

Covey, R. A. (2006). *How the Incas built their heartland: State formation and the innovation of imperial strategies in the Sacred Valley, Peru*. Ann Arbor: University of Michigan Press.

Cruikshank, J. (1985). The gravel magnet: Some social impacts of the Alaska highway on Yukon Indians. In K. Coates (Ed.), *The Alaska highway: Papers of the 40th anniversary symposium* (pp. 172–187). Vancouver: University of British Columbia Press.

Dahal, R. K., Hasegawa, S., Bhandary N. P., & Yatabe R. (2010). Low-cost road for the development of Nepal and its engineering geological consequences. In Geologically Active Williams et al. (Eds.), *IAEG 2010 Conference*. London: Taylor & Francis Group. Retrieved from http://www.hils.org.np/ranjan/papers/Ranjan_Low-cost_road.pdf.

Flower, J. (2004). A road is made: Roads, temples, and historical memory in Ya'an county, Sichuan. *The Journal of Asian Studies, 63*(3), 649–685.

Folke, C., Colding, J., & Berkes, F. (2002). Synthesis: Building resilience for adaptive capacity in social-ecological systems. In F. Berkes, J. Colding, & C. Folke (Eds.), *Navigating social-ecological systems: Building resilience for complexity and change* (pp. 352–387). Cambridge, UK: Cambridge University Press.

Forman, R., Sperling, D., Bissonette, J., Clevenger, A., Cutshall, D., Fahrig, L., France, R., Goldman, C., Heanue, K., Jones, J., Swanson, F., Turrentine, T., & Winter, T. (2002). *Road ecology*. Washington, DC: Island Press.

Gilio, B. (1986). Warfare in polonoroeste. *Cultural Survival Quarterly, 10*(2), 37. Retrieved from http://www.culturalsurvival.org/publications/cultural-survival-quarterly/brazil/warfare-polonoroeste.

Government of Nepal, Ministry of Federal Affairs and Local Development, Department of Local Infrastructure Development and Agricultural Roads (GoN/MoFAaLD/DoLIDAR). (2016). *Statistics of Local Road Network (SLRN) 2016*. Retrieved from https://www.dolidar.gov.np/.

Government of Nepal, Ministry of Finance (GoN/MoF). (2016). *Economic survey fiscal year 2015/16*. Retrieved from http://www.mof.gov.np.

Government of Nepal, Ministry of Physical Infrastructure & Transport, Department of Roads (GoN/MoPIT/DoR). (2016). *Statistics of Strategic Road Network (SSRN) 2015/2016*. Retrieved from http://dor.gov.np.

Government of Nepal, Ministry of Physical Planning and Works, Department of Roads (GoN/ MPPW/DoR). (2008). *Statistics of strategic road network SSRN 2006/07*. Kathmandu: HMIS Unit DOR.

Government of Nepal, Ministry of Physical Infrastructure and Transport, Department of Roads, Road Statistics(GoN/MoPIT/DOR). (2013–2014).

Grant, M. (1978). *History of Rome*. New York: Scribner's.

Hammarlund, A. (2010). The amber road: Center and periphery. *Baltic Worlds, 3*(1), 4–10.

His Majesty's Government of Nepal, Ministry of Local Development, Department of Local Infrastructure and Agricultural Roads (HMGN/MLD/DoLIDAR). (2005). *Integrated rural accessibility planning: Guideline*. Kathmandu: DoLIDAR. Retrieved from http://www.dolidar. gov.np/publication/publication.php.

His Majesty's Government of Nepal Ministry of Physical Planning and Works (HMGN/MPPW). (2001). National transport policy.

His Majesty's Government of Nepal Ministry of Works and Transport, Department of Roads (HMGN/MWT/DOR). (1985). *Nepal road statistics*. Kathmandu: Katunja Press.

Jackson, J. B. (1980). *The necessity for ruins, and other topics*. Amherst: University of Massachusetts Press.

Jacoby, H. G. (2000). Access to markets and the benefits of rural roads. *The Economic Journal, 110*, 713–737.

Kassam, K.-A. (2010a). Pluralism, resilience, and the ecology of survival: Case studies from the Pamir Mountains of Afghanistan. *Ecology and Society, 15*(2), 8.

Kassam, K.-A. (2010b). Coupled socio-cultural and ecological systems at the margins: Arctic and alpine cases. *Frontiers of Earth Science in China, 4*(1), 89–98.

Klatzel, F. (2000). *Green Roads: Building environmentally friendly, low maintenance rural roads through local participation*. Best Practices Report. GTZ Food for Work (Kathmandu GTZ).

Kreutzman, H. (2000). Improving accessibility for mountain development: Role of transport networks and urban settlements. In M. Banskota, T. S. Papola, & J. Richter (Eds.), *Growth, poverty alleviation and sustainable resource management in the mountain areas of South Asia. Proceedings of the International Conference, 31 January–4 February 2000* (pp. 485–513). Retrieved from http://www.mountainforum.org/rs/ol/searchft.cfm?step=vd&docid=711.

Lassoie, J., & Sherman, R. (2010). Promoting a coupled human and natural systems approach to addressing conservation in complex mountainous landscapes of Northwest Yunnan, China. *Frontiers of Earth Science in China, 4*, 67–82.

Laurence, R. (1999). *The roads of roman Italy: Mobility and cultural change*. London: Routledge.

Lay, M. G. (1992). *Ways of the word: A history of the world's roads and of the vehicles that used them*. New Brunswick, NJ: Rutgers University Press.

Liu, J. G., Dietz, T., Carpenter, S. R., Alberti, M., Folke, C., Moran, E., Pell, A. N., Deadman, P., Kratz, T., Lubchenco, J., Ostrom, E., Ouyang, Z., Provencher, W., Redman, C. L., Schneider, S. H., & Taylor, W. W. (2007a). Complexity of coupled human and natural systems. *Science, 317*(5844), 1513–1516.

Liu, J. G., Dietz, T., Carpenter, S. R., Alberti, M., Folke, C., Moran, E., Pell, A. N., Deadman, P., Kratz, T., Lubchenco, J., Ostrom, E., Ouyang, Z., Provencher, W., Redman, C. L., Schneider, S. H., & Taylor, W. (2007b). Supporting materials for: Complexity of coupled human and natural systems. *Science, 317*(5844), 1513–1516.

Liu, J. G., Dietz, T., Carpenter, S. R., Folke, C., Alberti, M., Redman, C. L., Schneider, S. H., Ostrom, E., Pell, A. N., Lubchenco, J., Taylor, W. W., Ouyang, Z., Deadman, P., Kratz, T., & Provencher, W. (2007c). Coupled human and natural systems. *Ambio: A Journal of the Human Environment, 36*(8), 639–649.

Liu, J. G., Dietz, T., Carpenter, S. R., Alberti, M., Folke, C., Moran, E., Pell, A. N., Deadman, P., Kratz, T., Lubchenco, J., Ostrom, E., Ouyang, Z., Provencher, W., Redman, C. L., Schneider, S. H., Liu, J., Linderman, M., Ouyang, Z., An, L., Yang, J., & Zhang, H. (2001). Ecological degradation in protected areas: The case of Wolong nature reserve for giant pandas. *Science, 292*(5514), 98–101.

Machlis, G. E., Force, J. E., & Burch, W. R. Jr (2005). *The structure of human ecosystems, V.05.2.* Retrieved from http://www.webpages.uidaho.edu/~gmachlis/files/model.pdf.

Machlis, G. E., Force, J. E., & Burch Jr., W. R. (1997). The human ecosystem part I: The human ecosystem as an organizing concept in ecosystem management. *Society and Natural Resources, 10*(4), 347–367.

McPeak, J., Lee, D., & Barrett, C. (2006). The dynamics of coupled human and natural systems. *Environment and Development Economics, 11*(1), 9–13.

Moseley, M. E. (2001). *The Incas and their ancestors: The archaeology of Peru* (Rev. ed.). London: Thames & Hudson.

Mulmi, A. D. (2009). Green road approach in rural road construction for sustainable development of Nepal. *Journal of Sustainable Development, 2*(3), 149.

Munguira, M. L., & Thomas, J. A. (1992). Use of road verges roads by butterfly and burnet populations, and the effect roads on adult dispersal and mortality. *Journal of Applied Ecology, 29*, 316–329.

Needham, J., Yates, R. D. S., & Wang, L. (1954). Science and civilisation in China. Cambridge [Eng.]: University Press.

Olsson, P., Folke, C., & Berkes, F. (2004). Adaptive co-management for building resilience in social-ecological systems. *Environmental Management, 34*(1), 75–90.

Oxley, D. J., Fenton, M. B., & Carmody, G. R. (1974). The effects of roads on populations of small mammals. *Journal of Applied Ecology, 11*, 51–59.

Pande, K. (2006). *Country status: Nepal.* Regional Experiences and Lessons in Financing Infrastructure and Improving Road safety 8–10 May 2006. Expert Group Meeting on the Development of the Asian Highway Network. United Nations Economic and Social Commission for Asia and Pacific Bangkok, Thailand. Retrieved from http://www.unescap.org/ttdw/roadsafety/StatusPapers2006/Nepal_statuspaper.pdf.

Paudyal, D. P. (1998). *Access improvement and sustainable development—Rural road development in Nepal.* Kathmandu: ICIMOD. Retrieved from http://lib.icimod.org/record/23957/files/Access%20Imporovement%20and%20Sustainable%20Development.pdf.

Pohl, W. (n.d.). *Trekking the Annapurna Circuit in Nepal: Change on the Horizon.* Retrieved from http://www.gonomad.com/1100-trekking-in-nepal-change-on-the-horizon#hmjrFLR2yf1QCRag.99.

Price, D. P. (1989). *Before the bulldozer: The Nambiquara Indians and the World Bank.* Maryland: Seven Locks Press.

Raj, P. A. (1978). *Road to the Chinese border.* Kathmandu: Foreign Affairs Journal Publication.

Richards, P. (1984). The economic context of rural roads. In J. Howe & P. Richards (Eds.), *Rural roads and poverty alleviation.* Boulder, CO: Westview Press.

Salick, J., Yongping, Y., & Amend, A. (2005). Tibetan land use and change near Khawa Karpo, eastern Himalayas. *Economic Botany, 59*(4), 312–325.

Schreiber, H. (1961). *The history of roads: From amber route to motorway.* London: Barrie and Rockliff.

Shah, N. (2007). Transport sector performance and impact indicators—A Nepalese case study. World Bank Report.

Shrestha, A. (2011). *Art Since the Time of Arniko.* ECS Nepal. Retrieved May 2011, from http://ecs.com.np/features/art-since-the-time-of-arniko.

Shrestha, B. (1981). *An introduction to Nepalese economy.* Kathmandu: Ratna Pustak.

Shrestha, R. L. (1980). *Impact of Kathmandu Raxaul highway (Tribhuban Rajmarg) on Nepalese economy (1956–1975).* Kathmandu: Katunja Press.

Sitaula, T. P. (2009). *Infrastructure development in Nepal: Opportunities and challenges for engineers.* Retrieved from http://scaef.org.np/.

Sushil Koirala wins vote to be Nepal's prime minister. (2014, January 10) British Broadcasting Corporation. Retrieved from http://www.bbc.com/news/world-asia-26116387.

Thapa, A. J. (2013). *Status paper on road safety in Nepal.* Europe-Asia Road Safety Forum and The 67th Session of the Working Party 1 (WP 1) of UNECE. New Delhi, India, 4–6 December 2013. Retrieved from http://www.dor.gov.np/documents/Status_Paper%20_2013.pdf.

Total road length crosses 80,000km. (2015, July 12), *The Himalayan.* Retrieved from https://thehimalayantimes.com/business/total-road-length-crosses-80000km/.

Ubukawa, T., de Sherbinin, A., Onsrud, H., Nelson, A., Payne, K., Cottray, O., & Maron, M. (2014). A review of roads data development methodologies. *Data Science Journal, 13*, 45–66. http://dx.doi.org/10.2481/dsj.14-001.

Ullman, E. (1956). The role of transportation and the bases of interaction. In J. Thomas (Ed.), *Man's changing role in changing the face of the earth* (pp. 862–880). Chicago: The University of Chicago Press.

United Nations Development Program (UNDP). (2006). *Developing rural transport and infrastructure. Millennium development goals needs assessment for Nepal.* Retrieved from http://www.undp.org.np/publication/html/mdg_NAN/Chapter_9.pdf.

Vehicles grew 15 times in 20 yrs. (2011, April 22). *The Himalayan.* Retrieved from http://www.thehimalayantimes.com/fullTodays.php?headline=Vehicles+grew+15+times+in+20+yrs&NewsID=285120.

Walsh, S., & McGinnis, D. (2008). Biocomplexity in coupled human-natural systems: The study of population and environment interactions. *Geoforum, 39*, 773–775.

World Bank (WB). (2006). *Infrastructure: Lessons from the last two decades of World Bank Engagement* (Discussion Paper). Retrieved May 28, 2010, from http://www-wds.worldbank.org.

Part II
Mobility as a Social and Ecological System

Chapter 2
Mobilities

Abstract Starting in the 1990s social scientists began to question some basic assumptions within their discipline. Chief among these was the widespread notion that sedentarism reflected normal human behavior. The common ground for those interested in mobilities and the reason the *Mobilities Turn* is different from other approaches is the emphasis to begin from a perspective of mobility rather than sedentary, of fluidity rather than fixity, and of variable rather than circumscribed. The sedentarist moves in a space that has been strictly appropriated, each person has their own enclosed parcel. The greater space is necessarily fragmented by these parcels enclosed by walls, fences, and boundaries and is further fragmented by the roads that connect these spaces. The nomad moves in a space that is open, smooth, and indeterminate with roads and paths that connect the open spaces rather than delimit and close the space.

The *Mobilities Turn* has its foundations in the work of several influential individuals whose writings pushed the boundaries of sedentarism such as Marc Augé (1995), Manuel Castells (1996), and James Clifford (1997). In addition Caren Kaplan (1996) was a catalyst for looking at mobility through a gender lens. Gendered mobility is a field of study that investigates the ways in which mobilities and gender intersect and "how mobilities enables/disables/modifies gendered practices". In the Nepal Himalaya mobility is still largely limited by the lack of well built and maintained roads. Gendered mobility patterns are embedded in specific Hindu caste group and tribal cultural traditions and vary widely dependent on very context specific factors as well as individual family dynamics.

Keywords Mobilities Turn • Gendered mobility • Sedentarism • Nomadism • Migration • Sense of place • Imaginary spaces • Coupled social and ecological systems • Transhumance

Mobility is defined as "the movement of people from one place to another in the course of everyday life … the daily rounds of activities such as paid and unpaid work, leisure, socializing and shopping" (Hanson 2010: 7). However, it is not just a physical

© The Author(s) 2017
R.E. Beazley, J.P. Lassoie, *Himalayan Mobilities*, SpringerBriefs
in Environmental Science, DOI 10.1007/978-3-319-55757-1_2

phenomenon; it is also social as well. It is socially produced and unlike movement, it is a contextualized process. Starting in the 1990s social scientists began to question some basic assumptions within their discipline. Chief among these was the widespread notion that sedentarism reflected normal human behavior. The common ground for those interested in mobilities and the reason the *Mobilities Turn* is different from other approaches is the emphasis to begin from a perspective of mobility rather than sedentary, of fluidity rather than fixity, and of variable rather than circumscribed.

Part II provides an overview of mobilities and the *Mobilities Turn* as well as how mobilities embrace coupled social and ecological systems. Finally, it considers mobilities in the context of communities in the Nepalese Himalayas.

2.1   The *Mobilities Turn*

The *Mobilities Turn* differentiates itself from previous geographies of inquiry by taking motion, a basic constituent of life, as its nucleus (Cresswell 2010). Urry (2007: 18) argues that due to this different focus it elucidates hidden or subterranean "theories, methods and exemplars of research...." that previously were not seen. This focus on motion as a lens to view life creates a framework that intersects other fields of research that formerly were bounded and allows for the creation of new synergies that produce transformative theoretical and methodological vistas. In addition, the *Mobilities Turn* recognizes coupled social and ecological systems and the interplay between the physical structures of mobility (roads, bridges, vehicles, etc.) and the aesthetics of the human experience, thereby linking bio/physical sciences, the humanities, and social sciences. This view argues that mobilities are more than just functional, that the parts that make up mobilities include such entities as design, implementation, governance, meaning, politics, livelihoods, and ethics. Another difference is that mobilities encompass multiple scales of movement and forms, from walking to global economic and labor movements. This includes the way in which these forces interact and influence each other in relation to movement, and the politics of movement. Finally, it reemphasizes the need to transcend boundedness and a sedentary framework (Cresswell 2010).

In contrast, sedentarism was proximately derived from the Hiedeggerian idea of dwelling places, bounded spaces where people dwell, reside, feel at home, and peaceful.

> Circumspect heedfulness decides about the nearness and farness of what is initially at hand in the surrounding world. Whatever this heedfulness dwells in from the beginning is what is nearest, and regulates our de-distancing. (Heidegger 1953: 100)

This concept infers sedentariness and its associations with short distance, static locale, stability, and fixed sense of place, as normal.

> As being-in-the-world, Da-sein essentially dwells in de-distancing. This de-distancing, the farness from itself of what is at hand, is something that Da-sein can *never cross over.* (Heidegger 1953: 100 emphases in the original)

The opposite then, long distance, on the move, and no one fixed sense of place or multiple sense of place is thought to be abnormal. (Sheller and Urry 2006).

In not-staying, curiosity makes sure of the constant possibility of *distraction*. Curiosity has nothing to do with the contemplation that wonders at being, *thaumazein,* it has no interest in wondering to the point of not understanding. Rather, it makes sure of knowing, but just in order to have known. The two factors constitutive for curiosity, *not-staying* in the surrounding world taken care of and *distraction* by new possibilities, are the basis of the third essential characteristic of this phenomenon, which we call *never dwelling anywhere.* Curiosity is everywhere and nowhere. This mode of being-in-the-world reveals a new kind of being of everyday *Da·sein,* one in which it constantly uproots itself. (Sheller and Urry 2006: 161)

Sedentarism necessarily posits human identity in circumscribed place, bounded territories of scale from household, to community, region, and nation, and therefore a distinct entity for social science inquiry. This sedentary view of human identity can be traced back to the Enlightenment.

Space is believed to be a zone of freedom. But like all aspects of freedom after the European Enlightenment, that zone is structured by property relations and contests between states and corporations for dominance and wealth. (Kaplan 2006: 400)

Sedentarism became further embedded in human identity during the rise of the high modernist state directed application of scientific and technical innovations creating a vision of a new Western society and later by the flurry of nation building following World War II (Scott 1998, 2009). Social sciences fixation on sedentarism is perhaps best exemplified by its lack of recognition of the significance of one of the crowning achievements of high modernism, the automobile and the change in mobility it inspired (Sheller and Urry 2000).

The *Mobilities Turn* has its foundations in the work of several individuals whose writings pushed the boundaries of sedentarism such as Marc Augé's (1995) *Non-places: Introduction to an Anthropology of Supermodernity*, Manuel Castells' (1996) *The Rise of the Network Society*, and James Cliffords' (1997) *Routes: Travel and Translation in the Later Twentieth Century.* In addition, Caren Kaplan's (1996) *Questions of Travel: Postmodern Discourses of Displacement* was a catalyst for looking at mobility through a gender lens. At the turn of the century condensed temporal spatial realities brought about by the components of globalization coalesced into a framework centered on mobilities as elucidated in John Urry's (2000) *Sociology Beyond Societies: Mobilities for the Twenty-first Century* and Victor Kaufmann's (2002) *Re-thinking Mobility: Contemporary Sociology* (Cresswell 2010). Urry's (2000) work examines the traditional concepts of society through the lens of state control, boundaries delimited by the state that control mobilities, to generate the concept what he calls "sociology beyond societies" (Urry 2000: 5).

In the final chapter an agenda for a *sociology beyond societies* is developed, organised around the distinction between gardening and gamekeeping metaphors. The emergence of gamekeeping involves reconsidering the nature of a civil society of mobilities; seeing how states increasingly function as 'regulators' of such mobilities; dissolving the 'gardening' distinction between nature and society; and examining the emergent global level that is comprised of roaming, intersecting, complex hybrids. (Urry 2000: 5)

Table 2.1 Categories of mobility (Adapted from Urry 2007)

Mobility	Characteristic	Examples/technologies
Capable	Being capable of movement, or having the property of movement, something that moves or facilitates movement	Mobile: people, phones, homes, hospitals, physical prostheses, societal class mobility as in the "new mobility" Makimoto and Manners (1997)
Mob	Unruly, disorderly, no fixed boundaries, hard to regulate and requiring new forms of governance	Unruly: mobs of people, multitudes, smart mobs/flash mobs Rheingold (2002)
Social	Social, vertical as in upward and downward social mobility, hierarchal, entangled interrelationships of physical motion and social mobility	The new Chinese billionaires, the rise of a middle class in India and other examples of the rising "virtual middle class" Friedman (2013) such as in Egypt
Migration	Horizontal semi-permanent geographical movement of people, often involves movement to another country or continent due to economic, climatic, political, issues of conflict or other reasons	Environmental migrants Brown (1976), migration for work, political migration (political refugees), diasporas, sex, food, and security Idyorough (2015)

Urry (2007) delineates four distinct categories of mobilities (Table 2.1). He argues that social interactions posit various aspects of the movement and/or fixity of people, concepts, items, and forms, therefore different modes of mobility create different types of societies. In addition, he points out that there is a credo/ideology of movement as a basic human right enshrined in the United Nations Universal Declaration of Human Rights (UDHR) and European Union's Charter of Fundamental Rights of the European Union (CFREU) as follows:

> (1) Everyone has the right to freedom of movement and residence within the borders of each state. (2) Everyone has the right to leave any country, including his own, and to return to his country. (UN UDHR 1948 Article 13)

In the United States, we may have an assumed right of the freedom of movement, but there is no provision in the US Constitution or other US laws that explicitly gives its citizens the right to the freedom of movement (Wilhelm 2010).

In academic scholarship, the conceptual bookends of mobility are sedentary metaphysics and nomadic metaphysics. Sedentary metaphysics looks at the world in terms of a rootedness in a specific place.

> This is a sedentarism that is peculiarly enabling of the elaboration and consolidation of a national geography that reaffirms the segmentation of the world into prismatic, mutually exclusive units of "world order" (Smith 1986:5). This is also a sedentarism that is taken for granted to such an extent that it is nearly invisible. And, finally, this is a sedentarism that is deeply metaphysical and deeply moral, sinking "peoples" and "cultures" into "national soils," and the "family of nations" into Mother Earth. It is this transnational
> Cultural context that makes intelligible the linkages between contemporary celebratory internationalisms and environmentalisms. (Malkki 1992: 31)

James Scott in *Seeing Like A State* (1998) delves into this type of sedentary metaphysics seen as a tool used by the state to make citizens more legible and therefore easier to control. In contrast, nomadic metaphysics sees the world in terms of a multiplicity of roots, which are in an on-going state of change and with no fixed point of origin. Deleuze and Guattari (1988: 330) further elucidate nomadic metaphysics.

> The nomad has a territory; he follows customary paths; he goes from one point to another; he is not ignorant of points (water points, dwelling points, assembly points, etc.). But the question is what in nomad life is a principle and what is only a consequence. To begin with, although the points determine paths, they are strictly subordinated to the paths they determine, the reverse of what happens with the sedentary. The water point is reached only in order to be left behind; every point is a relay and exists only as a relay. A path is always between two points, but the in-between has taken on all the consistency and enjoys both an autonomy and a direction of its own. The life of the nomad is the intermezzo. Even the elements of his dwelling are conceived in terms of the trajectory that is forever mobilizing them.

The sedentarist moves in a space that has been strictly appropriated, each person has their own enclosed parcel. The greater space is necessarily fragmented by these parcels enclosed by walls, fences, and boundaries and is further fragmented by the roads that connect these spaces. The nomad moves in a space that is open, smooth, and indeterminate with roads and paths that connect the open spaces rather than delimit and close the space (Deleuze and Guattari 1988). The discourse around globalization borrows freely from nomadic metaphysics putting a positive spin in the name of progress on all things mobile. This is in contrast to the current negative discourse about illegal and undocumented immigrants and current xenophobia toward Islam in the United States. It is possible to hold both views about different spaces at the same time without contradiction; for example, those in favor of closed borders for immigration but open borders for trade. However, rarely are either metaphysics found in absolute form, usually falling somewhere on a continuum between the two (Cresswell 2006).

Nonetheless, metaphors about mobility have become the ubiquitous neo-modern language just as the cell phone has become both the literal and metaphorical symbol. Take for example, the latest advertisement from Bavarian Motor Works (BMW).

BMW i Concept:
   BMW i stands for visionary electric cars and mobility services, inspiring design and a new understanding of premium that is strongly defined by sustainability. With BMW i the BMW Group is adopting an all-embracing approach, redefining the understanding of personal mobility with purpose-built vehicle concepts, a focus on sustainability throughout the value chain and a range of complementary mobility services.

The Future of Urban Mobility:
   "Spring 2008. A top-secret location somewhere in Munich, Germany. BMW's most innovative thinkers. And one ambitious goal: the radical reinvention of individual mobility in megacities."[1]

The problem with using mobility as a metaphor as well as the twentieth century's most potent male symbol of individual mobility, the car, is that it decontextualizes the gendered nature of mobility (Gasner 2009). In response to this misappropriation, feminists have criticized its use, pointing out the basic flaw in this essentialist take on mobility and travel. Caren Kaplan (1996: 401) contextualizes this flaw, "Light, sight,

---

[1] http://www.bmw-i-usa.com/en_us/concept/.

and travel become structuring concepts for this European Enlightenment subject, a subject that is arguably generically masculine, raced, propertied, and individualised in a legal as well as political, psychological, and philosophical sense."

2.2 Gendered Mobility

Gender is constructed through enacted reiteration within specific cultural temporal/ spatial geographies nested in different scales of magnitude. It is embedded in culture and influences access and use of transport.

> 'Access' is primarily a gendered phenomenon in the developing countries, pertaining to all the subsets of access, i.e. access to information, rights, land, money, education, skills, political participation and voice. (Uteng 2011: 1)

Gendered mobility is a field of study that investigates the ways in which mobilities and gender intersect and "how mobilities enables/disables/modifies gendered practices" (Uteng and Cresswell 2008: 1). When we consider gendered mobility, we should pay particular attention to both physical and social context. There are many contexts, which influence gendered mobility including:

- Urban or rural setting
- Cultural context and cultural norms
- Socio-economic class status
- White women or women of color
- Ethnic identity
- Level of education

The evolution of gendered mobility in relation to automated transport must entail consideration of boats, railways, public buses, and automobiles. Each of these form of transportation have their own unique gendered mobility characteristics. This section will focus only on the use of public buses and automobiles because they best suit the purpose of this book.

In England and the United States, the use of the automobile was initially considered too difficult for a woman. However, over time women showed that they could operate automobiles just as well as men. Nonetheless, within the context of women that could and did operate automobiles initially only privileged white women of higher socio-economic status made up this group (Ganser 2009). However, due to a number of factors including both women's rights and civil rights movements, the structure of American urban, suburban, and rural residence and lifestyle, and the availability of credit this situation has changed in the United States, presently more women than men have driver's licenses (More Women Have Driver's Licenses 2012). At the other end of the spectrum, there are places such as Saudi Arabia, where culturally dictated norms of gendered mobility are designed to keep women in very limited gendered spaces.

> There are no specific traffic laws that make it illegal for women to drive in Saudi Arabia. However, religious edicts are often interpreted as prohibiting female drivers. Such edicts also prevent women from opening bank accounts, obtaining passports or even going to school without the presence of a male guardian. (Jamjoon 2013)

Transgressing these male guardian system edicts can result in a jail sentence or in some cases a beating.

In May 2011, Manal al-Sharif, a single mother and technology consultant, posted a You Tube video of herself driving with a female friend around Al Khobar in eastern Saudi Arabia talking about women's rights.[2] In one day, her video received 700,000 hits. She was arrested and held for nine days charged with "inciting women to drive"; her only crime was driving her own car. Even though the video was blocked within her country it was the catalyst for a movement called Women2Drive that petitioned the government for the right of Saudi women to drive. On June 17, 2011, Women2Drive and Manal al-Sharif organized the *I Will Drive My Own Car* campaign. More than 100 Saudi women drove their cars along with their male supporters. Shaima Jastania was one of the women who tested this system by driving a car in Jeedah, for which she received a sentence of 10 lashes after being brought before a judge. Fortunately, for Shaima, King Abdullah overturned the sentence (Al Omran 2011). Since then Manal al-Sharif has received death threats which has led many of her family members to leave the country, she was pressured to resign from her job which meant she also lost her housing (McVeigh 2012). Ironically Foreign Policy magazine named her one of the "Top 100 Global Thinkers" in 2011 ("The FP Top 100 Global Thinkers" 2011) and Time included her in their list of the "100 most influential people of 2012" (Baker 2012). Al-Sharif was also nominated for the 15th annual Vital Voices Global Leadership Awards at the Kennedy Center, which she was not able to attend because, "…Saudi officials, angered by an earlier al-Sharif trip to the United States, made it clear to her, as one source put it, 'that there would be consequences for her family if she chose to go to Washington'" (Fineman 2012). Since women are considered minors in Saudi Arabia, when they leave the country they must have a male guardian sign a document at the border or airport. There is now a Short Message Service tracking system that alerts male guardians if a woman is about to leave the country (Paramaguru 2012). Things are slowly changing however, in 2012 Saudi Arabia for the first time sent a women's team to the Olympics and the Saudi religious police recently lifted a ban on women using bikes (Quan 2013).

Gendered mobility and gendered space is deeply engrained in many mythological forms that are embedded in culture. For example, *Little Red Riding Hood* (Grimm et al. 1900) is a fairy tale that has been read to countless children and has its equivalent in other cultures as well. If we consider what takes place in this story with a gender lens, we can see that it is a tale about "normative spatial behavior dictated by prevalent gender roles; this is reflected in a didacticism suggesting that girls need to safeguard themselves when they step out of the home" (Ganser 2009: 13). While we can appreciate the adventurousness of Little Red Riding Hood for

---

[2] http://www.youtube.com/watch?v=sowNSH_W2r0.

stepping out of the house alone in the face of convention and going against her mother's wishes by taking an alternate route, we also see that she pays for that transgression (Cixous 1981). After arriving at her grandmother's house, she gets in bed with her grandmother (the wolf in disguise) and is eaten. In the Brothers' Grimm version, a woodcutter (male) intercedes and slays the wolf releasing Little Red Riding Hood and her grandmother from its belly, an interesting twist that reinforces the normative gendered space. As Ganser (2009: 13) observes:

> In any case, the tale is an early example of a road narrative that familiarizes the protagonist not only with the effects of the patriarchal spatial and symbolic orders on the gendered individual as well as with the resulting limitations of agency, but also, if the tale is re-read from a feminist perspective like Cixous', with the potential pleasures of transgressive mobility.

While in most fairy tale books, Little Red Riding Hood is depicted as a white girl; the struggle for equity in gendered space was even harder for women of color in the United States. They face not only the normative masculine constructs of gendered mobility, but also the race-based restrictions imposed by the white population at large.

> Nineteenth century black women were more aware of sexist oppression than any other female group in American society has ever been. Not only were they the female group most victimized by sexist discrimination and sexist oppression, their powerlessness was such that resistance on their part could rarely take the form of organized collective action. (Hooks 1981: 161)

This gendered and 'racialized' mobility applied not just to African Americans but also to other groups in the U. S., including East Indian (Bald 1983), Chinese (Lee 2000), Japanese (Twomey 2009), Hispanic (López 2008), and Native American (Hirschfelder 1995).

The roots of gendered mobility can be traced much further back than the story of *Little Red Riding Hood*. A recent study by Montano et al. (2013), emphasizes how gendered mobility is embedded and shaped in patrilineal cultures. "In accordance with the prevalence of patrilocal habits, where women move to their husbands households after the marriage, higher female transgenerational migration rates have been inferred at both local and continental level in most populations studied (Montano et al. (2013): 14–18)." These finding supports an earlier study on African populations by Seielstad et al. (1998) entitled *Genetic evidence for a higher female migration rate in humans*. As Marilyn Strathern (1988: 82) notes in *The Gender of the Gift*, in Melanesian society;

> The modelling of a social life on notions of male rootedness or the mobility that has displaced it, and on their superior organizational skills, could be taken as an ideological set fashioned by men to further their own interests, including the domination of women.

If we consider developing countries, only the rich are able to afford cars and the poor are restricted to the cheapest form of transport, which usually involves riding in an overcrowded bus, jeep, or minibus. Gendered mobility, however, is more than just about vehicles. Ultimately, gendered mobility affects livelihoods through gendered division of labor. This is not surprising considering that women provide 60–80% of the food in developing countries and over half of the world's food supply, in addition to doing most of the household chores such as water, fodder, and fuel wood collection, cooking, child and parental care, and cleaning (Das 1995). Cunha (2006: 8) sums up this reality in Africa:

> In sub-Saharan Africa, women account for 70% of household labour and 85% of household daily effort spent on transport. They carry at least three times more ton/kilometres per year than men. They walk between 15–30 hours per week on transport-related chores, carrying between 30–50 kilograms and frequently with a baby on their backs. These are heavier loads than the maximum 20 kg recommended by the International Labour Organisation, and commonly result in long-term health problems. To help their mothers, young girls are often removed from school to assist with chores (Heyen-Perschon 2001; Peters 2001; Omar 2001; Starkey 2001; World Bank 2002).

Studies of gender and transportation in developing countries have brought to light some interesting sociocultural aspects of road construction. Fernando and Porter (2002), after compiling many case studies from around the world concerning transportation and gender, concluded that culture has a very strong influence on women's freedom to use transport services. In most cases, cultural rules allow men to travel without restriction provided they have the funds. Women however, do not always have the same opportunity to benefit from improved or new roads when a road reaches their community. Access to funds within households and work responsibilities inside and outside the house can preclude women from taking advantage of transport options in the same way that men do. For example, in some areas of Tanzania sociocultural traditions dictate that only men should own and use transport devices. Consequently, women carry loads on their head instead of using bicycles on the road (Mwankusye 2002).

In some cases, such as in Saudi Arabia as mentioned above, religious practices affect whether women can take advantage of new road construction and consequent mobility benefits. One case study in a Muslim area of Nigeria found that the religious tradition of female seclusion precluded the ability of women to take full advantage of road improvement and better transport, because they rarely travel outside the village (Yunusa et al. 2002). Another interesting case is from a Muslim area of Bangladesh where even though the village maintained the tradition of female seclusion, poorer women were not as restricted as richer women due to the cultural stigma of travel not being respectable for women of higher socioeconomic status (Matin et al. 2002).

Some of the benefits cited include better access to schools and health facilities and increased participation of men in chores often falling to primarily women, such as collecting firewood and accompanying children to school and the sick to a hospital (Yunusa et al. 2002).

2.2.1 Gendered Mobility in Nepal

Seddon and Shrestha (2002), in their study area in Nepal found that road access in some cases increased the workloads of both women and children, because male members of the family would travel to markets and seek off-farm employment, leaving them to take up the additional work burden. These authors also commented on how improvements in roads and transportation have facilitated the commercial sex trade. Women can make three times the amount of money in the sex trade as they can from labor, and for many women it provided a way to ensure adequate income for their family's survival. The town of Siliguri, on the southern border of Nepal, has become the main hub of girl trafficking and prostitution in Nepal because it has the best road links to India, where most of the girls are sold into prostitution ("Siliguri Link to Nepal Sex Trade" 2009). An International Labor Organization report estimated that 12,000 children are trafficked from Nepal every year (Bal Kumar et al. 2001).

Ghimire (2002) found a similar increase in the workloads of women associated with road construction. He found that improved road access in a mountainous area of Nepal increased the demand for milk and dairy products, resulting in more cattle rearing. The extra burden of rearing more cattle including collecting fodder and mulch has traditionally been women's work. With the increase in cattle, their workload also increased. However, in the Terai where the topography is flatter he found that women's burden decreased due to the road because they were able to use bicycles to travel and carry small items. In addition, the collection of firewood, which used to be a responsibility assumed only by women, was changing because men could now use bikes to transport firewood. Nevertheless, he points out that with the increased mobility afforded by roads, additional economic activities become available and as a result more travel and transport are required and women are typically the ones who shoulder this additional work.

Sylvain and Devries (2012) assert that Tamang women they studied in Dhidopur, Southeast of Kathmandu, were constrained in their movement for labor migration mainly due to do social norms. If a woman leaves the village to seek labor her behavior can no longer be monitored by her kin and neighbors, which can lead to gossip and tarnish her family's reputation. In particular, this had to do with both the potential to be lured into sex trafficking and concerns for her personal safety. The authors argue that these views are heightened by the rhetoric about sex trafficking that is prevalent in Nepal and that it discourages the women in her research site from even trying to travel. In the authors' opinion, the money spent on these "scare tactics" could be better used to provide more education for women about sex trafficking and for developing safer transport for women such as women only modes.

In a 2011 paper, Craig argues that migration, in this case from Mustang, Nepal to New York, fuels the imagination and creates realms of possibility not only for the migrant but also for the friends and family who receive their stories at home in Mustang. The case study she examines concerned the imagination of health care in New York as opposed to Nepal. "Out-migration from Mustang has created different demands on bodies, alternate ways of understanding causes and conditions of illness, new patterns of seeking medical care" (Craig 2011: 210). Ultimately, there was no one answer, all the migrants had different experiences, from those who thought western medicine was better to those who thought traditional medicine was better and some a combination of the two. The significance in terms of mobility is that migration can shape the imagination in different ways. In the case of the Tamang women in Dhidpour, it had negative connotations that translated into socially constructed prescriptions against women travelling. In the case of Mustang, the conclusion is less clear because the results of migration were interpreted as both positive and negative. What is clear is that imagination is a space within which there is mobility to imagine the ideal or the problematic. In the physical realm, the individual conceives of particular locations as "… specific places on the map and idealized spaces in which people imagine and enact certain possibilities for living, and experiences of suffering." (Craig 2011: 193).

> For example, illness experiences and subjectivities can be altered through migration such that karma (T. *lé*), as a source of illness causality or divination (T. *mo*), as a way of determining illness causality and an appropriate treatment regime, are no longer viable explanations or practices. (Craig 2011: 194)

Craig further argues that migration can shape ideas of personhood and subject as well as the spaces of possibility. These possibility spaces can be interpreted as manifestations of sense of place embedded with life experiences both positive and negative. Physical and metaphorical possibility spaces can affect individual health regimes.

> People from Mustang sometimes connect specific diseases to patterns of migration, both enduring and novel: from the lowland fevers (T. ring gyi tsawa) of seasonal moves from high-altitude homelands to lower elevations in Nepal or India, to "tension" or "depression" as idioms of distress that take on particular meanings in New York or Kathmandu. In contrast, people describe a sort of biopsychosocial peace with longing and nostalgia when they speak of Mustang's air (T. *lung*, N. *hawa*) and water (T. *chu*, N. *paani*) and how it agrees with them—the possibilities for living it engenders, and their compromised well being when in other places. (Craig 2011: 195)

Imaginary spaces interact in physical spaces, either far removed from the imagined space or, due to migration, in the imagined space. When these combinations of spaces intersect, there can be a physical manifestation of the psychological space. In either case, the mobility within these spaces provides the agency, or lack of, for the evolution of the process. In this regard, we can see how mobility in the physical world is intimately connected and shaped by mobility in the imaginary world, and that both are gendered spaces with gendered mobility.

Much has been written about the restriction of mobility placed on women in Nepal by the Hindu caste system (Kondos 2004; Rankin 2003; Bennett 1983; Agarwal 1994; Cameron 1998) where women are expected to remain close to home and only go out while accompanied by a male relative, adhere to gendered taboos, shoulder the domestic chores, and prioritize the male family members over females. This restricted mobility reduces the female household members chances of attending school past a basic level due to the socially constructed expectation that girls will ultimately be householders when they grow up and therefore there is no need for them to attend school when they can be actively working at home (Onta and Resurreccion 2011; Rothchild 2006; Maslak 2003).

Tamang (2000) speaks to some of the core issues of gendered mobility in Nepal by taking a historical political view of the state institutionalization of patriarchy in Nepal. In analysing the changes to family law in Nepal's civil code, the *Muluki Ain* instituted in 1854, she argues that during the Panchyat years (1961–1990) "…the state and the law played central roles in the structuring of a particular form of patriarchy—a shift from "family patriarchy" to "state patriarchy" (Tamang 2000: 127). While Tamang makes no claims as to existent affects this change had on specific caste and ethnic groups in Nepal, she does recognize the larger overall impact of furthering the social construction of gendered space vis a vis the reproductive roles of women being confined to private space while the productive roles of men occupy public space.

Finally, several other studies have been done on gender and transport in Kathmandu (Neupane and Chesney-Lind 2014; Action Aid 2013; Harrison 2012; Udas 2012; Paudel 2011; ADB 2010a, b). These studies highlight the daily harassment women in Kathmandu face as they try to negotiate moving in the city. The overcrowded public transit systems leads to groping and other form of sexual harassment, which is endemic in Kathmandu. The lack of proper street lighting makes it dangerous for women to approach bus stops at night and the poor quality of the sidewalks, or in many cases lack of sidewalks, can often cause physical injury. While there have been several groups who have tried to make gender and transport issues more public in Kathmandu such as "Safe City Nepal",[3] "Social Service Awareness Raising and Advocacy for Tranquility and Humanity",[4] the international organization "Meet Us on the Streets"[5], and events such as Stop Violence Against Women[6] progress has been slow. Nonetheless, change is happening. Sajha Yatayat, a newly reopened bus company (May 2013), is hiring women conductors to sell tickets on the buses. Their buses have restrictions for only 15 standing passengers and they have installed closed circuit TV cameras as well. Namita, one of the female conductors, thinks Sajha buses are safer.

> Sajha bus is a legacy and I think people appreciate the service. It is also safer for us women to work because not more than 15 people can stand and there's a lot of space. I think people respect Sajha and wouldn't ever think of vandalizing the seats or spitting, like it happens in other vehicles. Being a woman, I cannot imagine working as a conductor in other public transports, except Sajha. With the way so many people crammed in, there's hardly any space for the conductor to stand. (Tripathi 2013)

2.3 Conclusions

Gendered mobility is a fundamental aspect of daily life in both developed and developing countries. Most of the inquiry into gendered mobility has focused on travel patterns of women in developed countries (Fahs 2011; Ganser 2009; Wilke 2007; Morgan 2001; Rosenbloom 1993; Hamilton et al. 1991; Jones 1990; Grieco et al. 1989;

[3] https://www.facebook.com/safecitiesnepal.

[4] http://www.saathnepal.org/.

[5] http://www.meetusonthestreet.org/.

[6] https://www.facebook.com/safecitiesnepal/events.

Little et al. 1988). These studies were important to establish that men and women have very different travel patterns, namely that men usually make fewer trips of longer distance but more direct with very few stops, whereas women make many shorter distance trips but these trips involve many stops and many different routes and therefore often take more time then men's trips. The different nature of the trips requires different types of access to different modes of transport, especially for the more complicated travel patterns of women. Nonetheless, the studies of women's travel patterns in developed countries while useful in looking at developing countries, does not have the depth of complexity needed to understand gendered mobility in radically different cultural and geographical spaces such as rural mountain villages in the Nepal Himalaya. While there have been advances in Nepal in terms of inclusion of gender mainstreaming in most development projects, the reality is that not much has changed not only in Nepal but worldwide with gender mainstreaming in development projects. "Despite extensive discourse and resources that focus on women as key actors for development, their situation has not changed considerably (Cunha 2006: 4)". In terms of the Gender Equity Index (GEI) Nepal falls towards the bottom with a score of 47 with Afghanistan at the very bottom with a score of 15 and Norway at the top with a score of 89 (Social Watch 2012).

There is a pressing need for a better understanding of gendered mobility in developing countries. To do this requires site-specific information that will help in designing transport schemes that are gender equitable. Barwell (1996: 26–27) has identified five elements that are crucial in understanding the scope of transport load that is placed on women in rural locations:

- Number of female adults in the household;
- Distance to sources of water and firewood;
- Number of children in the household that can help, especially daughters;
- Food preferences of the household—this is related to the distance needed to travel to process different staples such as rice or flour;
- Availability and use of individual means of transport

This information could serve as a baseline for a more targeted study to determine what transport modes would work best in each individual setting. Additional indicators that reflect the availability and access to transport, such as maternal mortality rates as an indicator of how easy it is to get to medical centers, could be incorporated into a feasibility study and would help planners tailor transport modes for specific contexts.

In the above example, this type of approach will also help in planning health facility locations and in addition adapt individual means of transport that provide the broadest range of access. Uteng (2009: 69) notes: "Given such benchmarking, it will become easy to assess the specific kinds of alterations needed in the mobility systems to adapt towards gendered needs (for example, usage of mobile phones to substitute the missed trips and access information)."

These are just a few examples of how to begin the process of assessing and adapting mobilities toward gender equity. One of the key elements is changing planners' traditional narrow sighted concentration on connecting point A to B to one of looking at the "lived in" environment of the space where mobility is needed (Gudmundsson et al. 2005). It is in this "lived in" space that the subtleties of gendered mobility can be recognized and addressed. Mobility is about moving in spaces to accomplish daily activities. To understand mobility needs, we must therefore understand the daily activities in the space where people live and how that space is socially structured and gender prescribed. Another key element that has been lacking is the need to bridge the gap between the two main research lenses that have been used in gendered mobilities research thus far. One approach focuses on how gender influences mobility; the other investigates how mobility affects gender. As Susan Hanson (2010: 5) suggests:

> I argue for the need to shift the research agenda so that future research will synthesize these two strands of thinking along three lines: (1) across ways of thinking about gender and mobility, (2) across quantitative and qualitative approaches, and (3) across places. In the final part of the essay I suggest how to achieve this synthesis by making geographic, social and cultural context central to our analyses.

2.4   Mobilities in the Context of Coupled Social and Ecological Systems

Human mobility transpires in multidimensional spaces. Geophysical, geopolitical, cultural, spiritual, social, community, family, personal, gendered, and psychological spaces are a few examples of different and often overlapping spaces we move through and interact in. While geophysical space is often first thought of when considering mobility, other socially constructed spaces influence how we move through geophysical space. The fact that a road has been constructed into a previously road less area does not guarantee equal access and ease of mobility along the road. Environmental and social factors can both facilitate and limit who and what is able to move along the road. In Nepal for example, the spread of transport syndicates/cartels affect the mobility of people who wish to travel on roads in both urban and mountain environments (Syndicate in Transportation, Again 2013). These syndicates control transport operators in given areas requiring passengers to use only vehicles operated by the syndicate (Pokharel and Gautam 2014). Non-syndicate operated vehicles have been stopped and barred entry to areas controlled by syndicates, essentially giving syndicates a monopoly on transport in the areas where they exert control (Transporters Misbehave With Tourists 2014). This allows the syndicates to control prices often charging more for transport than the market would allow if there was healthy competition (Sharma 2011). In this way we can see how road construction affects mobility as a coupled social and ecological system. The physical/ecological system of building a hardened earth surface to increase the mobility of people and goods is intimately coupled to the social systems through which the road traverses and affects the relative mobility of those wishing to take advantage of the new potential for increased mobility.

2.5 Himalayan Mobilities

Traditionally Himalayan mountain people have had to migrate seasonally to survive. Transhumance is one example of the importance of mobility as an adaptive livelihood strategy in mountainous regions. In many areas of the Himalayas winter provides the impetus for migration both to escape to warmer southern climates and to engage in small scale trade to supplement income during the slack winter agricultural season. In Nepal, before the last decade, there was migration to the hills from both the south and the north. From the south it was motivated by a desire to escape the heat and malarial disease of the plains, and find arable land. From the north it was to escape harsh environmental conditions and seek better agricultural land. As the population continued to increase in the hills, the carrying capacity was reached in terms of cultivatable land. As result, for the last hundred years, there has been a general migration from the hills to the plains (Hrabovszky and Miyan 1987). This has increased even more since the late 1950s, when DDT was used to eradicate malarial mosquitoes in the Terai (Gill 2003).

Domestic seasonal migration is nothing new to farmers in the hills and mountains of Nepal, as they have been practicing transhumance for hundreds of years, an adaptive strategy that accommodates the seasons and the ecological restoration of the seasonal grazing areas. In the mountains the amount of cultivatable land is much less than in the hills and Terai, and the harsh climate and short growing season makes agriculture alone a very hard way to make a living. Many mountainous areas have both food deficits and high levels of poverty. This combination of poverty and food insecurity is the 'push' factor in the seasonal migration in Nepal, while the availability of labor in southern areas and India is the 'pull' factor. Additional alternative livelihood strategies have been pursued for centuries in Nepal to fill the gap left by seasonal agriculture. This migration is advantageous in a number of ways. It reduces the number of family members that need to be fed on the farm (the push factor) and when the migrants return just before summer to begin the agricultural cycle again, they often bring food supplies with them, and if they were fortunate some savings (the pull factor) (Brown 2008; Bohle and Adhikari 1998).

Two other the main adaptive strategies, which are interrelated, are migration for temporary or seasonal job opportunities, and trade. In Nepal trade frequently involves temporary and often seasonal migration. This migration has taken many different forms, including the Trans Himalayan Salt Trade, and the shift in the salt trade to the south after the Tibetan border was closed in 1959. Employment as soldiers in the Gurkha Division of Indian and British armies is a long-standing tradition for Gurung and other ethnic groups since the nineteenth century (Stringer 2011; Gurung 2004). Migration to serve in the armed forces historically is the first large-scale migration of men from the hills in Nepal (Gill 2003). Similar trends in migration and mercenary military employment have been documented in other mountainous areas of Asia, such as the Karakoram (Kreutzman 1991). In the late 1980s to early 1990s, the carpet-manufacturing boom stimulated a huge migration trend to the Kathmandu Valley (Graner 2001). A more recent trend that is having a large impact on Nepali society is the outmigration for jobs in the Gulf countries and Malaysia. This physical mobility has enabled outmigrated Nepalis to send remittances that have in turn enhanced socioeconomic mobility and dramatically increased the size of Nepal's middle class over the last two to three decades (ADB 2010b; Liechty 2003). Remittances have been credited with reducing poverty and contributing to between 25 and 33% of Nepal's GDP (WB 2014; Jones and Basnett 2013).

The coupled social and ecological nature of mobility is evident not only in the creation of a new middle class but also in terms of the social mobility afforded to previously marginalized groups such as *Dalits* (occupational caste groups), who in some cases have been able to rise above their previously disadvantaged and socially immobile status through the injection of remittances from relatives abroad (Gautam 2014). This speaks to the multi-dimensional spaces of mobilities mentioned previously and stresses the importance of recognizing the complex connectivity inherent in the social and ecological couplings of mobilities.

References

Action Aid International. (2013). *Women and the city II: Combating violence against women and girls in urban public spaces*. The Hague, The Netherlands: Action Aid International.

Agarwal, B. (1994). *A field of one's own: Gender and land rights in South Asia*. Cambridge, UK: Cambridge University Press.

Al Omran, A. (2011, September 19). *Reports: Saudi king cancels lashing sentence against woman who drove, NPR*. Retrieved from http://www.npr.org/blogs/thetwo-way/2011/09/28/140887141/reports-saudi-king-cancels-lashing-sentence-against-woman-driver.

Asian Development Bank (ADB). (2010a). *Gender equality and social inclusion action plan Kathmandu Sustainable Urban Transport Project(RRPNEP44058-01)*.Retrievedfromhttp://www.adb.org/projects/documents/kathmandu-sustainable-urban-transport-project-gender-equality-and-social-inclusio.

Asian Development Bank (ADB). (2010b). *Key indicators for Asia and the Pacific 2010, Special Chapter The Rise of Asia's Middle Class*. Asian Development Bank. Retrieved from http://www.adb.org/publications/key-indicators-asia-and-pacific-2010.

Augé, M. (1995). *Non-places: Introduction to an anthropology of supermodernity*. London: Verso.

Baker, A. (2012). The 2012 TIME 100: The 100 most influential people in the world. *Time, 179*(17).

Bal Kumar, K. C., et al. (2001). *Nepal trafficking in girls with special reference to prostitution—A rapid assessment*. International Labor Organization International Programme on the Elimination of Child Labour (IPEC). Retrieved from http://www.ilo.int/legacy/english/regions/asro/newdelhi/ipec/download/resources/nepal/nppubl01eng9.pdf.

Bald, S. R. (1983). *From Satyartha Prakash to Manushi: An overview of the "women's movement" in India*. East Lansing: Women in International Development, Michigan State University.

Barwell, I. (1996). *Transport and the village, findings from African village-level travel and transport surveys and related studies*. Sub-Saharan Africa Transport Policy Program, The World Bank and Economic Commission for Africa (SSATP Working Paper No. 23).

Bennett, L. (1983). *Dangerous wives and sacred sisters: Social and symbolic roles of high-caste women in Nepal*. New York: Columbia University Press.

Bohle, H., & Adhikari, J. (1998). Rural livelihoods at risk how Nepalese farmers cope with food insecurity. *Mountain Research and Development, 18*(4), 321–332.

Brown, L. (1976). *World population trends: Signs of hope, signs of stress* (Worldwatch Paper 8). Washington, DC: Worldwatch Institute.

Brown, O. (2008). *Migration and climate change*. Geneva: International Organization for Migration. Retrieved from http://www.iom.cz/files/Migration_and_Climate_Change_-_IOM_Migration_Research_Series_No_31.pdf.

Cameron, M. (1998). *On the edge of the auspicious: Gender and caste in Nepal.* Urbana: University of Illinois Press.

Castells, M. (1996). *The rise of the network society.* Malden, MA: Blackwell Publishers.

Cixous, H. (1981). 'Castration or decapitation?' (Annette Kuhn, Trans.). *Signs, 7*(1), 19–55.

Clifford, J. (1997). *Routes: Travel and translation in the late twentieth century.* Cambridge, MA: Harvard University Press.

Craig, S. (2011). Migration, social change, health, and the realm of the possible: Women's stories between Nepal and New York. *Anthropology and Humanism, 36*(2), 193–214.

Cresswell, T. (2010). Towards a politics of mobility. *Environment and Planning D: Society & Space, 28*(1), 17–31.

Cresswell, T. (2006). *On the move: Mobility in the modern western world.* New York: Routledge.

Cunha, C. (2006). *Bicycles as a lever for rural women's empowerment lessons learned from bicycle projects in sub-Saharan Africa and an alternative approach in southern Mozambique, Conference sub-theme: Gender, transport and economic development.* Retrieved from http://www.benbikes.org.za/namibia/pdfs/Bicycles%20and%20empowerment%20Paper.pdf.

Das, M. (1995). *Improving the relevance and effectiveness of agricultural extension activities for women farmers.* Rome: Food and Agriculture Organization of the United Nations.

Deleuze, G., & Guattari, F. (1988). *A thousand plateaus: Capitalism and schizophrenia.* London: Athlone.

EU CFREU. (2007). *Charter of fundamental rights of the European Union, Title V Citizen's Rights, Article 45: Freedom of movement and residence.* Retrieved from http://eur-lex.europa.eu/en/treaties/dat/32007X1214/htm/C2007303EN.01000101.htm.

Fahs, A. (2011). *Out on assignment: Newspaper women and the making of modern public space.* Chapel Hill: University of North Carolina Press.

Fernando, P., & Porter, G. (Eds.). (2002). *Balancing the load: Women, gender, and transport.* New York: Zed Books.

Fineman, H. (2012, June 7). *Manal Al-Sharif, Saudi Right-To-Drive Protester, Skips U.S. Award To Avoid 'Consequences', Huffington Post.* Retrieved from http://www.huffingtonpost.com/2012/06/07/manal-al-sharif-saudi-protester_n_1576454.html.

Friedman, T. (2013, February 2). The virtual middle class rises. *The New York Times.* Retrieved from http://www.nytimes.com/2013/02/03/opinion/sunday/friedman-the-virtual-middle-class-rises.html.

Ganser, A. (2009). *Roads of her own: Gendered space and mobility in American women's road narratives, 1970–2000.* Amsterdam: Rodopi.

Gautam, H. (2014, June 13). Foreign employment lifts Dalit families out of poverty. *Republica.* Retrieved from http://www.myrepublica.com/portal/index.php/thweek/slc/ads/ncell.swf?action=news_details&news_id=78889.

Ghimire, G. (2002). Transport in the mountains and the Terai: Kushiya Damrang and Shivpur, Nepal. In P. Fernando & G. Porter (Eds.), *Balancing the load: Women, gender, and transportation* (pp. 246–257). New York: Zed Books.

Gill, G. (2003). *Seasonal labour migration in rural Nepal: A preliminary overview* (Working paper 218). Overseas Development Institute. Retrieved from http://www.odi.org.uk/resources/download/1783.pdf.

Graner, E. (2001). Labor markets and migration in Nepal. *Mountain Research and Development, 21*(3), 253–259.

Grieco, M., Pickup, L., & Whipp, R. (1989). *Gender, transport and employment.* Aldershot: Avebury Press.

Grimm, J., Grimm, W., & Pocock, N. (1900). *Grimm's fairy tales.* New York: Doran.

Gudmundsson, H. et al. (2005). Mobility, sustainability and beyond. In T. U. Thomsen, L. Drewes & H. Gudmundsson (Eds.), *Social perspectives on mobility*, epilogue, Aldershot: Ashgate.

Gurung, H. (2004). *Mountain reflections.* Kathmandu: Mandala.

Hamilton, K., Jenkins, L., & Gregory, A. (1991). *Women and transport: Bus deregulation in West Yorkshire.* Bradford: University of Bradford.

Hanson, S. (2010). Gender and mobility: New approaches for informing sustainability. *Gender, Place & Culture, 17*(1), 5–23.

Harrison, J. (2012). *Gender segregation on public transport in South Asia: A critical evaluation of approaches for addressing harassment against women.* M.Sc. Dissertation, School of Oriental and African Studies, University of London.

Heidegger, M. (1953). *Being and time.* Oxford: Blackwell.

Heyen-Perschon, J. (2001). *Non-motorised transport and its socio-economic impact on poor households in Africa: Cost-benefit analysis of bicycle ownership in rural Uganda.* Uganda: FABIO/BSPW.

Hirschfelder, A. B. (1995). *Native heritage: Personal accounts by American Indians, 1790 to the present.* New York, NY: Macmillan.

Hooks, B. (1981). *Ain't I a woman: Black women and feminism.* Boston, MA: South End Press.

Hrabovszky, J., & Miyan, K. (1987). Population, growth, and land use in Nepal "the great turnabout". *Mountain Research and Development, 7*(3), 264–270. Proceedings of the Mohonk Mountain Conference: The Himalaya-Ganges Problem, 264–270. Retrieved from http://www.jstor.org/stable/3673203.

Idyorough, A. (2015). *Sociological analysis of social change in contemporary Africa.* Makurdi: Aboki Publishers.

Jamjoon M. (2013, April 15) Billionaire Saudi prince tweets support for women driving. *CNN.* Retrieved from http://edition.cnn.com/2013/04/15/world/meast/saudi-prince-women-driving.

Jones, P. (Ed.). (1990). *Developments in dynamic and activity-based approaches to travel analysis.* Aldershot: Avebury Press.

Jones, H., & Basnett, Y. (2013). *Foreign employment and inclusive growth in Nepal.* Overseas Development Institute. Retrieved from http://www.odi.org/publications/7424-foreign-employment-inclusive-growth-nepal-can-be-done-improve-impacts-people-country.

Kaplan, C. (2006). Mobility and war: The cosmic view of US 'air power'. *Environment & Planning A, 38*(2), 395–408.

Kaplan, C. (1996). *Questions of travel: Postmodern discourses of displacement.* Durham, NC: Duke University Press.

Kaufmann, V. (2002). *Re-thinking mobility: Contemporary sociology.* Aldershot: Ashgate.

Kondos, V. (2004). *On the ethos of Hindu women: Issues, taboos, and forms of expression.* Kathmandu: Mandala Publications.

Kreutzman, H. (1991). The Karakoram highway: The impact of road construction on mountain societies. *Modern Asian Studies, 25*(4), 711–736.

Lee, E. E. (2000). *Nurturing success: Successful women of color and their daughters.* Westport, CT: Praeger.

Liechty, M. (2003). *Suitably modern: Making middle-class culture in a new consumer society.* Princeton, NJ: Princeton University Press.

Little, J., Peake, L., & Richardson, P. (1988). *Women in cities—Gender and the urban environment.* New York: New York University Press.

López, I. (2008). "but you don't look Puerto Rican": The moderating effect of ethnic identity on the relation between skin color and self-esteem among Puerto Rican women. *Cultural Diversity & Ethnic Minority Psychology, 14*(2), 102–108.

Makimoto, T., & Manners, D. (1997). *Digital nomad.* Chichester: John Wiley.

Malkki, L. (1992). National geographic: The rooting of peoples and the territorialization of national identity among scholars and refugees. *Cultural Anthropology, 7*(1), 24–44.

Maslak, M. A. (2003). *Daughters of the Tharu: Gender, ethnicity, religion, and the education of Nepali girls.* New York: RoutledgeFalmer.

Matin, N., Mukib, M., Begun, H., & Khanam, K. (2002). Women's empowerment and physical mobility implications for developing rural transport, Bangladesh. In P. Fernando & G. Porter (Eds.), *Balancing the load: Women, gender, and transport* (pp. 128–150). New York: Zed Books.

McVeigh, T. (2012, June 16) Saudi Arabian women risk arrest as they defy ban on driving. *The Guradian.* Retrieved from http://www.guardian.co.uk/world/2012/jun/17/saudi-arabian-women-risk-arrest-ban-driving.

Montano, V., Marcari, V., Pavanello, M., Anyaele, O., Comas, D., Destro-Bisol, G., & Batini, C. (2013). The influence of habitats on female mobility in central and western Africa inferred from human mitochondrial variation. *BMC Evolutionary Biology, 13*(24), 1–9.

More women have driver's licenses than men in United States for the first time ever. (2012, November 12). *Associated Press New York daily news.* Retrieved from http://www.nydaily-news.com/autos/women-driver-licenses-men-article-1.1200847#ixzz2Tln2OklG.

Morgan, M. (2001). *National identities and travel in Victorian Britain.* Houndmills: Palgrave.

Mwankusye, J. (2002). Do intermediate means of transport reach rural women? In P. Fernando & G. Porter (Eds.), *Balancing the load: Women, gender, and transportation* (pp. 37–49). New York: Zed Books.

Neupane, G., & Chesney-Lind, M. (2014). Violence against women on public transport in Nepal: Sexual harassment and the spatial expression of male privilege. *International Journal of Comparative and Applied Criminal Justice, 38*(1), 23–38.

Omar, M. (2001). Gender mobility: The long and winding road to women's transportation solutions. *The Economist.* Retrieved from http://www.economist.com/node/820453.

Onta, N., & Resurreccion, B. (2011). The role of gender and caste in climate adaptation strategies in Nepal. *Mountain Research and Development, 31*(4), 351–356.

Paramaguru, K. (2012, November 24). Your wife has just left the country: Saudi Arabia implements SMS-tracking system. *Time.* Retrieved from http://newsfeed.time.com/2012/11/24/your-wife-has-just-left-the-country-saudi-arabia-implements-sms-tracking-system/#ixzz2Tl4CILHH.

Paudel, R. (2011). *A research report on understanding masculinities in public transport.* Kathmandu: Nepal submitted to SANAM Fellowship Program. Retrieved from http://www.engagingmen.net/files/resources/2012/sysop/Nepal_Radha_A_Research_Report_on_Understanding_Masculinities_in_Public_Transport_Kathmandu_Nepal.pdf.

Quan, K. (2013, April 3). *Saudi women can now ride bicycles in public (kind of), time.* Retrieved from http://world.time.com/2013/04/03/saudi-women-can-now-ride-bicycles-in-public-kind-of/#ixzz2Tl5QNWJu.

Peters, D. (2001). *Gender and transport in less developed countries: A background paper in preparation for CSD-9.* Retrieved from http://www.tuberlin.de/~isr/fachgebiete/planungstheorie/download/cv%20deike%20peters%20english.pdf.

Pokharel, S., & Gautam, H. (2014, March 23). Transport syndicate bars Indian tourists from visiting Muktinath. *Republica.* Retrieved from http://www.myrepublica.com.

Rankin, K. (2003). Cultures of economies: Gender and socio-spatial change in Nepal. *Gender, Place and Culture, 10*(1), 111–129.

Rheingold, H. (2002). *Smart mobs. The next social revolution.* Cambridge, MA: Basic Books.

Rosenbloom, S. (1993). Women's travel pattern at various stages of their lives. In M. Cindi Kand Janice (Ed.), *Full circles: Geographies of women over the life course* (pp. 208–242). London: Routledge.

Rothchild, J. (2006). *Gender trouble makers: Education and empowerment in Nepal.* New York: Routledge.

Scott, J. (2009). *The art of not being governed: An anarchist history of Upland Southeast Asia.* New Haven: Yale University Press.

Scott, J. (1998). *Seeing like a state: How certain schemes to improve the human condition have failed.* New Haven: Yale University Press.

Seddon, D., & Shrestha, A. (2002). Gender in rural transport development: Chattra Deruali, Nepal. In P. Fernando & G. Porter (Eds.), *Balancing the load: Women, gender, and transport* (pp. 235–245). New York: Zed Books.

Seielstad, M. T., Minch, E., & Cavalli-Sforza, L. (1998). Genetic evidence for a higher female migration rate in humans. *Nature Genetics, 20*(3), 278–280.

Sharma, S. (2011, April 27). Syndicates and cartels: The bane of Nepalese economy. *TFAS Asia.* Retrieved from http://tfasasia.wordpress.com.

Sheller, M., & Urry, J. (2006). The new mobilities paradigm. *Environment & Planning A, 38*(2), 207–226.

Sheller, M., & Urry, J. (2000). The city and the car. *International Journal of Urban and Regional Research, 24*, 737–757.

Smith, A. (1986). *The ethnic origins of nations*. New York: Blackwell.

Starkey, P. (2001). *Local transport solutions: People, paradoxes and progress. Lessons arising from the spread of intermediate means of transport*. SSATP Working Paper No. 56. Retrieved from http://www.animaltraction.org/Local-Transport-Solutions-RTTP-Starkey-Jan01-540kb.pdf.

Stringer, K. D. (2011). Global counterinsurgency and US army expansion: The case for recruiting foreign troops. *Small Wars & Insurgencies, 22*(01), 142–169. doi:10.1080/09592318.2011.546604.

Sylvain, R., & Devries, S. (2012). *Mobility Matters: Tamang Women's Gendered Experiences of Work, Labour Migration and Anti-Trafficking Discourses in Nepal*. Retrieved from http://hdl.handle.net/10214/3494

Syndicate in transportation, again. (2013, January 10). *Karobar National Economic Daily*. Retrieved from http://www.karobardaily.com.

Tamang, S. (2000). Legalizing state patriarchy in Nepal. *Studies in Nepali History and Society, 5*(1), 127–156.

"The FP Top 100 Global Thinkers". (2011). *Foreign Policy, 190*, 34–108.

Transporters misbehave with tourists in Mustang. (2014, January 4) *Himalayan News Service*. Retrieved from http://www.thehimalayantimes.com.

Tripathi, P. (2013, May 15). *Sajah Bus/Yatyat in Kathmandu Nepal*. Retrieved from http://sajhabus.blogspot.com/2013/05/women-conductors-at-sajha-breaking.html#.UZfBuqJwrSg.

Twomey, C. (2009). Double displacement: Western women's return home from Japanese internment in the second world war. *Gender & History, 21*(3), 670–684.

Udas S. (2012). *Public transport quality survey*. A report for Clean Air Network Nepal (CANN) and Clean Energy Nepal (CEN). Retrieved from http://www.cen.org.np/uploaded/Public%20Transport%20Survey%20report.pdf.

UN UDHR. (1948). United Nations. *The Universal Declaration of Human Rights*. Article 13. Retrieved from https://www.un.org/en/documents/udhr/.

Urry, J. (2007). *Mobilities*. London: Polity.

Urry, J. (2000). *Sociology beyond societies: Mobilities for twenty first century*. New York: Routledge.

Uteng, T. (2011). *Gender and mobility in the developing world*. World Development Report, Gender Equality and Development Background Paper, World Bank. Retrieved from http://siteresources.worldbank.org/INTWDR2012/Resources/7778105-1299699968583/7786210-1322671773271/uteng.pdf.

Uteng, T. (2009). Gender, ethnicity, and constrained mobility: Insights into the resultant social exclusion. *Environment & Planning A, 41*(5), 1055–1071.

Uteng, T., & Cresswell, T. (2008). *Gendered mobilities*. Aldershot, England: Ashgate.

Social Watch (2012) *The social watch report 2012: A global progress report on poverty eradication and gender equity*. Retrieved from http://www.socialwatch.org/report2012.

Wilhelm, K. (2010). Freedom of movement at a standstill? Toward the establishment of a fundamental right to intrastate travel. *Boston University Law Review, 90*(6), 2461–2494.

Wilke, J. (2007). *Eight women, two model Ts, and the American west*. Lincoln: University of Nebraska Press.

World Bank (WB). (2014). *Migration and development brief 22, migration and remittances: Recent developments and outlook*. Retrieved from http://siteresources.worldbank.org/INTPROSPECTS/Resources/334934-1288990760745/MigrationandDevelopmentBrief22.pdf.

World Bank. (2002). *Improving rural mobility: Options for developing motorized and non-motorized transport in rural areas*. World Bank Technical Paper No. 525. Retrieved from http://documents.worldbank.org/curated/en/819101468780344782/Improving-rural-mobility-options-for-developing-motorized-and-nonmotorized-transport-in-rural-areas.

Yunusa, M., Shaibu-Imodagbe, E., & Ambi, Y. (2002). Road rehabilitation: The impact on transport and accessibility. In P. Fernando & G. Porter (Eds.), *Balancing the load: Women, gender, and transport* (pp. 111–118). New York: Zed Books.

Part III
Challenges and Impacts of Building Roads in the Himalayas

Part III provides an in depth assessment of the positive and negative impacts of road development in Nepal on the environmental (Chap. 3), socioeconomic (Chap. 4), and sociocultural (Chap. 5) spheres, and how these are interrelated and influence each other. After providing a review of related work in Nepal and elsewhere, the core of each chapter will be the consideration of three specific case studies based on empirical research by the senior author in Nepal in April-May and October-November 2009 (Beazley 2013). This fieldwork was conducted along the Annapurna Circuit Trail (ACT) in the Annapurna Conservation Area in the west-central part of Nepal (Figure 3.0) to investigate the impacts of expanding road networks along the ACT on local communities. During the spring period research was conducted on the west side (see map) of the ACT in the Kali Ghandaki River valley (Case Study #1) (Figure 3.0). We refer to the road on the west side as the Kali Ghandaki Highway. Fall period research was conducted along the entire length of the ACT beginning in Besisahar and finishing at the west side terminus of Birenthanti (Case Study #2) (Figure 3.0). We refer to the east side road as the Marsyangdi Highway. Several side trips away from the ACT were made to investigate the extant of the road impact zone (Figure 3.0). On the east side this included parts of the Gurung Heritage Trail and the Nar Phu area. On the west side a side trip was made to the Annapurna Base Camp. Case Study #3 looks at two specific villages, one on the east and one on the west side of the trail, to highlight some unintended and unforeseen consequences of road building.

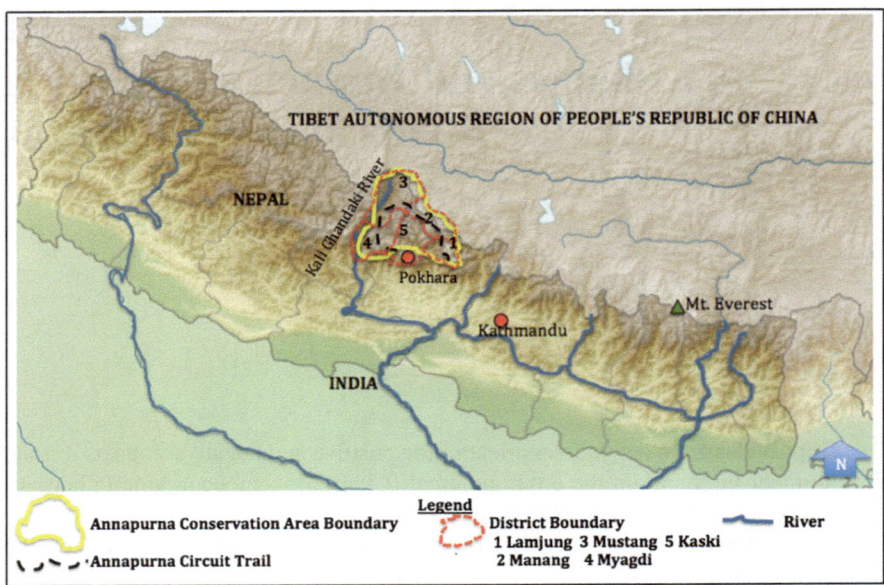

Figure 3.0 Annapurna Conservation Area Research Site (Adapted from http://www.map1001.com/Nepal+map-4.html)

Methods included structured household questionnaires, both semi-structured and open interviews, and photo documentation with the help of a Nepali field assistant/ translator. When possible, discussions arising from 2009 fieldwork are updated with information obtained during a more recent field study period (2014–2015) by the senior author. In the following text all participants' names have been changed to protect anonymity.

Chapter 3
Environmental Challenges and Impacts

Abstract Road construction involves a profound alteration of the earth's surface, which leads to many negative impacts on the environment. These impacts are frequently not just localized events restricted to the particular road location and often have far reaching effects in other areas. While some effects are immediate others evolve over a period of time and involve an interconnected chain of events. Consequently these effects can be categorized as direct, indirect, and cumulative impacts. To assess the overall environmental impact of road construction projects, activities associated with the project that do not take place at the main site should be considered also. For example, the materials used for the project may actually be excavated from a site some distance from the project area. The impacts can be grouped into several categories including land and water effects, biodiversity and habitat loss, pollution, and road kill. In Nepal one of the most common environmental disasters are landslides. In many places landslides are the direct result of road cuts and improper cut and fill techniques. This combined with intense seasonal monsoonal rains every year results in loss of life, livestock, valuable agricultural land, and homes and blocks mobility corridors. Many of these effects can be seen along the Marsyangdi Highway in Lamjung and Manang Districts. This road (recently completed in 2016) has had acute impacts on local communities and on the most popular trekking route in Nepal, the Annapurna Circuit Trail.

Keywords Environmental impacts • Landslides • Monsoon • Habitat fragmentation • Trekking • Annapurna Circuit Trail • Marsyangdi highway • Environmental disasters • Erosion • Road construction • Cut and fill techniques

This chapter addresses the environmental challenges and impacts of road construction in Nepal and the Himalayas, including different types of construction such as local construction and *Green Roads* and the coupled social and ecological systems elements that influence environmental impacts. Empirical research on the environmental impacts of the Kali Ghandaki Highway as well as the Marsyangdi Highway (Beazley 2013) is presented using photo documentation to highlight the narrative. In addition, case studies from other roads in Nepal are analysed in comparison to both of the above-mentioned highways.

© The Author(s) 2017
R.E. Beazley, J.P. Lassoie, *Himalayan Mobilities*, SpringerBriefs
in Environmental Science, DOI 10.1007/978-3-319-55757-1_3

3.1   Overview of Environmental Impacts of Roads

Road construction involves a profound alteration of the earth's surface, which leads to many negative impacts on the environment. These impacts are frequently not just localized events restricted to the particular road location and often have far reaching effects in other areas. While some effects are immediate others evolve over a period of time and involve an interconnected chain of events. Consequently these effects can be categorized as direct, indirect, and cumulative impacts. To assess the overall environmental impact of road construction projects, activities associated with the project that do not take place at the main site should be considered also. For example, the materials used for the project may actually be excavated from a site some distance from the project area.

3.1.1   Land and Water Effects

The most obvious initial effect is the alteration of the ground surface. This involves removing the surface cover and compaction of the soil beneath it, which can lead to a number of negative consequences. First, it decreases the amount of productive soil available for agriculture and depending on the road alignment it may actually displace a portion of land already under cultivation. Erosion often becomes a major problem as the vegetative root system holding the soil together is removed allowing water to easily carry it away often impacting areas far from the erosion area. Erosion can lead to many other serious problems such as slope instability and water quality issues that in turn affect plants, animals, and humans. Slope instability is further affected by a number of other factors; improper cut and fill techniques can lead to landslides as a result of a combination of making the cuts too steep, improper placement of the fill resulting in dangerously loaded slopes, and poor drainage (Tsunokawa and Hoban 1997). Road construction is cited frequently as a contributing factor in deforestation, erosion, and landslides in many countries (Petley et al. 2007; World Bank 2006; Daming et al. 2004; Ives 2004; LPSPT 2001).

3.1.2   Biodiversity and Habitat Loss

In terms of biodiversity the most detrimental impact of road construction is loss of habitat, habitat fragmentation, and edge effects (Laurance et al. 2007; Forman and Alexander 1998; Mader 1984). Furthermore, Wilcox and Murphy (1985: 884) state that most ecologist agree that "… habitat fragmentation is the most serious threat to biological diversity and is the primary cause of the present extinction crisis." Shaffer and Stein (2000) report that habitat loss is the greatest threat for 85% of the species identified as being endangered. It has been estimated in the United States that one mile of interstate highway eliminates up to 48 acres (19.4 ha) of habitat (CEQ 1976 cited in Noss 1995). A study in China's Yunnan province (Wang et al. 2007) showed that road network (RN) disturbance in the area resulted in a 1900 billion RMB (~US $300 billion) loss in eco-system services value (ESV) from 1985 to 2000. The authors predicted that as a result of increasing RN projects ESV would decrease by an average of 1.2% annually from 2010 to 2020, equalling an average loss of 30 billion RMB (~US$4.8 billion) per year in ESV. They further concluded that landscape fragmentation and eco-environment degra-dation was mainly the result of road construction networks.

Habitat loss can be the direct result of the road surface eliminating a piece of habitat or decreasing the overall size and accessibility through habitat fragmenta-tion. Fragmentation results from the barrier effect that roads create when they divide up unified pieces of landscape (Wei et al. 2010). This reduced connectivity has a profound effect on wildlife by restricting movement between local populations (Vos and Chardon 1998, Mader 1984; Oxley et al. 1974), which can result in a smaller population sizes. These isolated smaller populations are more at risk to extinction than larger populations due to several factors including the smaller gene pool size (Evink 2002). A study by Findlay and Houlahan in 1997 (cited in Findlay and Bourdages 2000: 87) found that "… for a sample of wetlands in south-eastern Ontario, current herptile, bird, and vascular plant species richness showed significant negative correlations with the (more or less contemporaneous) density of roads on adjacent lands, up to distances of at least 2 km from the wetland. From this and ancillary evidence, we inferred a causal relationship, namely, that road construction leads to reduced species richness because of one or more processes (e.g., restricted movement of species)." The introduction of invasive species is another result of road construction that contributes to loss of biodiversity (Lonsdale and Lane 1994; Cowie and Warner 1993). Invasive species can enter an area along roadways either by being planted to re-vegetate the roadside or are carried as seeds by vehicles from other areas. Habitat modifications arising from cutting, mowing, scraping, burning, and herbicide use in roadside maintenance can also negatively affect birds, other animals, and insects that inhabit these areas (Smith and Hellmund 1993).

3.1.3   Pollution

Pollution associated with roads includes air, land, water, light, and noise pollution. The use of vehicles on roads contributes a number of pollutants that contain heavy metals such as lead in diesel fuel, zinc in motor oil, and nickel in gasoline that end up in the soil in the vicinity of roads (White and Ernst 2003). Other chemicals asso-ciated with road maintenance such as de-icers, salt, and herbicides are common also (Amrhein et al. 1992). A report by the National Cooperative Highway Research Panel (Evink 2002: 19–20) states: "The EPA (1996) reports that between the mid-1980s to the mid-1990s approximately 10 million tons of rock salt was used on the nation's roads and caused at least 11% of the 20 impaired stream miles reported nationally." Many of these chemicals end up being transported to other areas by water either as surface runoff, in roadside ditches, or seeping into groundwater increasing water pollution levels and having negative impacts on plants and animals. Water quality can also be negatively affected when erosion caused by road construction adds high sediments loads to runoff. This can result in siltation and lower oxygen content affecting stream, lake, and wetland flora and fauna (Aruga et al. 2005).

Roads have a large impact on the hydrology of areas they pass through. Removing the surface vegetation affects not only the direction the runoff travels, but also the rate at which it percolates into the soil. Paved roads create an impervious surface that accentuates the diversion of surface runoff away from its natural pattern. White and Ernst (2003: 8) claim "A one-acre parking lot produces 16 times as much run-off as a one-acre meadow." Usually ditches are dug along roadsides to channel water off and away from roads and further alter the natural drainage pattern affecting groundwater recharge and channelling large volumes of water to new areas (Scheinder 2010). Many negative impacts can result, such as increased erosion and siltation at the point of ditch discharge and high concentrations of pollutants (White and Ernst 2003). Motor vehicles are responsible for a large portion of the gases that contribute to air pollution and create acid rain in addition to lowering visibility and air quality due to smog and dust. Noise pollution is another oblivious result of roads that has been shown to negatively affect breeding rates in birds (Reijnen et al. 1995), and road lights have been shown to affect both plants and animals (ERM 1996). These effects, in addition to the physical barrier created by roads, roadside ditches, and fences, alter the movement behavior of wildlife.

3.1.4 Road Kill

Vehicles are responsible for killing a large number of animals every year. The Humane Society and the Urban Wildlife Research Center estimated that one million vertebrates are killed every day across the United States' road network (Noss 1995). Forman and Alexander (1998: 212) state: "Sometime during the last three decades, roads with vehicles probably overtook hunting as the leading direct human cause of vertebrate mortality on land." There are several reasons why roads and vehicles have become such a deadly threat to wildlife. Roads cut across traditional migration routes and intersect home territories of many animals. As animals try to move along or across roads in relation to their natural habitat, death from vehicles is a frequent outcome. Wildlife may also attempt to cross a road to access sources of forage and water or may even find habitat suitable to them along the roadside (Harris and Scheck 1991). The impact of wildlife mortality by vehicles is highlighted by Harris and Schneck (1991: 192): "Collisions with motor vehicles is now known to be the number one source of mortality for all of Florida's large rare and endangered vertebrates including panther, black bear (*Ursus americanus*), key deer (*Odocoileus virginianus clavium*) and American crocodile (*Corocodylus acutus*)."

3.1.5 Positive Environmental Impacts

Only a few positive environmental impacts of roads have been reported, such as road verges providing habitat for birds, butterflies, and other small animals (Munguira and Thomas 1992; Adams 1984; Oxley et al. 1974). Roads in forested areas can also enhance growth by increasing available light and acting as firebreaks (Brown 2001). In addition, some roadside maintenance practices such as mowing and burning can be positive depending on the maintenance cycle and the type of vegetation that is present, but this requires careful planning not only for verges, but also for the construction of suitable wildlife movement corridors (Smith and Hellmund 1993).

3.2 Environmental Impacts in Nepal

Two of the major environmental impacts of road building in Nepal that have been well documented are erosion and landslides (Hasegawa et al. 2009; Dahal et al. 2008; Petley et al. 2007; Dahal et al. 2006; Merz et al. 2006; Merz et al. 2003; Hearn 2002; Upreti and Dhital 1996; Deoja 1994; Carson 1985). This relationship between roads, erosion, and landslides has been observed in other mountainous areas in Asia (Salick et al. 2005; Daming et al. 2004; Ives 2004; Hamiliton and Bruijnzeel 1997; Hewitt 1997; Deoja 1994; Karan and Iijima 1985) and other parts of the world (see Sidle et al. 2006). Landslides and floods are the most common natural disasters in Nepal. From 1983 to 2005 there were 6984 deaths caused by floods and landslides averaging 303 per year and equalling a third of all the fatalities from all natural disasters (Upreti 2006). The true figures are most likely substantially higher considering these are only the reported cases, and that the lack of communication networks in the rural mountainous areas of Nepal could mean that many deaths go unreported. From 1983 to 2003, landslides and floods destroyed 143,554 houses, averaging 7329 annually (Upreti 2006). Again, these figures are misleading because they only take into account the houses destroyed at the time of the disaster. Countless more sustain damage and destruction due to unstable land in the wake of the incident. The total cost of damage to infrastructure due to landslides and floods is approximately US$20 million per year in Nepal (Khanal 1996).

Roads are more influential in terms of the greatest landslide losses and surface erosion per unit area affected than any other type of land use. This type of mass wasting is significantly higher along roads then in steep forested areas that have not been disturbed (Sidle et al. 2006). In Nepal, Deoja (1994) estimated that 3000–9000 m^3 of landslides per kilometer of road occur during the construction of mountain roads. After completion, 400–700 m^3 of landslides per kilometer per year continues to occur along mountain roads. Road construction in Nepal results in significantly greater soil loss with estimates as high as 150 tons per hectare from poorly constructed roads (Deoja 1994).

A study east of Kathmandu in the Andheri Khola (river) catchment found that road construction had increased the sediment load to the river by 300–500% per year (Merz et al. 2006). After studying numerous landslides along roads in Nepal, Hagesawa et al. (2009: 1424) concluded that the combination of improper land use practices and seasonal monsoon rain make "… the Nepalese Himalayas the most unstable landscapes in the world." Their use of GIS mapping of landslides showed that road alignment had a significant relationship with landslide occurrence and that most of the Mugling-Narayanghat Road, the main road that accesses the Kathmandu Valley, runs through landslide-prone terrain. Dahal et al. (2006) reported 213 landslides along one section of the Mugling-Narayanghat Road that were triggered by rainfall. They go on to say that landslides in Nepal always occurred during the monsoon, implicating rainfall as the main triggering factor. It is not surprising then that improper drainage is cited frequently as one of the main problems with road construction in Nepal (Dahal et al. 2008; Dahal et al. 2006; Hearn 2002; Deoja 1994). In addition to poor drainage, cut and fill techniques can lead to slope instability if not done correctly.

During the construction of the Lamosangu-Jiri Road, cut mountain slope material was thrown down the valley side of the road destroying vegetation. Numerous landslides during the first monsoon occurred on both the mountainside of the road, due to undercutting the slope, and on the valley side, due to the loss of vegetation (Schaffner 1987). Road alignment is critical in terms of both avoiding unstable slopes and mass balancing of cut and fill materials. A poorly aligned road, and dumping fill material indiscriminately over the side, can lead to serious road problems that make maintenance very expensive and in some cases require realignment of the road (Schaffner 1987).

Hearn (2002) points out that many roads in mountainous areas such as Nepal are designed with insufficient data to assure proper alignment, thereby increasing landslide potential. He notes the need for an integrated approach involving geologic, soils, and geomorphological data. While this is the ideal situation, the reality in Nepal is that politics and budget constraints often preclude such an integrated approach. This is highlighted by the comment made by Dahal et al. (2006: 512), that the Department of Roads "… does not have a single geologist as an employee."

This frequent improper road assessment is exacerbated by the lack of uniform and compulsory national procedures, standards, and guidelines. Consequently, road construction is often dictated by the desires of politicians and donor agencies (Dahal et al. 2006). Even well planned and engineered roads in Nepal are not safe from landslides. The Dharan-Dhankuta Road, which was considered one of the best engineered roads at the time and cost over US$1 million per km to build, had to be closed periodically during monsoon season, including 1984 when mass wasting along the road created more than US$5 million in damage (Carson 1985). A study of landslide occurrence in Nepal by Petley et al. (2007) suggests that the recent trend in both landslide occurrence and landslide fatalities could be the result of the also increasing trend in rural road construction since the 1990s. He points out that these low budget projects may result in poor alignment and improper construction techniques, both of which have been shown to increase the likelihood of landslides along roads.

3.3 Summary

While not all road construction, road networks, and related vehicle effects have a negative impact on the environment, the large majority of them do contribute in some way to environmental degradation and loss of ecosystem services. The main impact is biodiversity loss arising from habitat fragmentation and degradation due to bisecting large contiguous patches of habitat, deforestation and erosion, slope instability and landslides, pollution from various road related activities, alteration of surface run-off and groundwater, and spread of invasive species. Forman and Alexander (1998) estimate that as much as 15–20% of the land surface area of the United States is affected by roads. Considering the amount of roads currently on the surface of the earth, it is obvious that roads represent one of the most negative anthropogenic impacts on the environment.

In Nepal, the major environmental impacts are soil erosion and landslides, which are often the result of poor alignment and improper construction techniques, including lack of balancing cut and fill materials and poorly designed drainage systems. However, even properly designed road projects are not immune to landslides due to the intense rainfall pattern during the monsoon period, which has been shown to be one of the main triggering factors for landslides in Nepal. There is a time lag in the interplay between the time when the road construction takes place in the dry season and the time when the heavy monsoon rains trigger landslides along the poorly constructed road cuts. In Nepal, landslides and other forms of mass-wasting such as debris flows constrain options for villagers livelihoods by removing productive agricultural land and crops, killing livestock, destroying houses, trails, and roads, closing roads and trails for months, and causing accidental deaths.

3.4 Case Study #1 Kali Ghandaki Highway

This case study illustrates how poorly planned and implemented road construction can lead to numerous environmental impacts including erosion, different forms of pollution, landslides and mass wasting, and uncontrolled spur road development (Beazley 2013). Due to the social and ecological systems couplings environmental effects create impacts in socioeconomic spheres. This will also be discussed.

3.4.1 Background

The Kali Ghandaki Highway is part of the North-South Transport Corridor Project, which aims to develop transportation corridors from Nepal's nine official northern border crossings with China to its southern border with India (WB 2005). In 2008, the Nepal Army completed construction of the Beni-Jomsom-Kagbeni Road

(Fig. 3.1), a major part of the northern portion of the Kali Ghandaki Highway. This road was built in sections, with the easiest sections built first, including an extension from Jomsom to Kagbeni, and a side road from Kagbeni to the sacred site at Muktinath (Kagbeni-Muktinath Road, F166). This whole section including the side road to Muktinath was built on the pre-existing Annapurna Circuit Trail (ACT), which used to be the most popular trekking destination in Nepal.

Construction of the road to connect Jomsom/Kagbeni north through Lo Manthang to Korala at the Chinese border (88 km) (Fig. 3.1) has been an on going project for some time. In 2001, a 20 km section from the Chinese border south to Lo Manthang was completed. It was started in 1999 financed by six Village Development Committees (VDCs) in Upper Mustang and was extended further south from Lo Manthang to Syangboche (Tashi Bista, owner Mustang trekking company and Upper Mustang resident, personal communication, June 27, 2014). There is some concern that this road will affect cultural sites in Upper Mustang. Most of the buildings in Upper Mustang use rammed earth construction methods. In Lo Manthang, the UNDP temporarily halted the construction of the road from the Tibetan border because the vibrations from vehicle noise near the rammed earth fortress walls around Lo Manthang and in the monasteries, threatened the integrity of these structures. After a thorough study, the UNDP changed the road alignment so vehicles would be further away and not affect the fortress walls and monasteries (Edwards et al. 2006).

Fig. 3.1 Kali Ghandaki Highway (*red*) with Muktinath spur road (*yellow*) (adapted from Google Earth)

Construction of a road leading from Kagbeni north to Lo Manthang has been slower due to the technical difficulties of the terrain and budget constraints. In the winter months when the Kali Ghandaki River is low, trucks and tractors often drive in the dry riverbed. When the river carries more water an alternative road was necessary to traverse the region from Kagbeni to Lomanthang, which was completed in 2014. According to the 2013 Mustang District Transport Master Plan (GoN/MoFAaLD/DoLIDAR 2013: 1) the entire road stretching from near Beni through Jomsom and Kagbeni to Lo Manthang and north to the Chinese border is called the Pairothapla-Jomsom-Ghoktang Road (F042). For simplicity it will be referred to as the Kali Ghandaki Highway in general except where certain sections required more specific names are needed. It is now (2016) possible to drive from Kagbeni all the way to the Tibetan borders with one change of vehicles where there is only a footbridge crossing the Kali Ghandaki River.

3.4.2   General Environmental Impacts

During the 2009 research period the noise and dust pollution from vehicles plying the road was amply evident (Figs. 3.2 and 3.3) as well as erosion and landslide scars (Fig. 3.4). Even though this road was built within the Annapurna Conservation Area Project (ACAP) the Nepal Army, which built the road, did not follow the mandatory road construction guidelines for ACAP (NTNC 2008, Binod Gurung, ACAP official, Jomsom, personal communication, November 29, 2009). The army is often used for road construction in Nepal.[1] It is doubtful whether the army has knowledge or experience of environmentally friendly road construction techniques and they did not follow the mandatory road construction guidelines for ACAP (NTNC 2008,

Fig. 3.2 Trekker and jeep on the Kali Ghandaki Highway (R. E. Beazley 2009)

---

[1] http://www.nepalarmy.mil.np/bpd.php.

Fig. 3.3 Trekker left in the dust (R. E. Beazley 2009)

Fig. 3.4 Kali Ghandaki Highway with landslide scar in the left foreground (R. E. Beazley 2009)

Binod Gurung, ACAP official, Jomsom, personal communication, November 29, 2009). From personal observation by the senior author in 2009, it appears that the local spur road building off the main Beni—Jomsom—Kagbeni Road in ACAP financed by the Village Development Committees (VDC) was being built the cheapest and quickest way possible. In addition, the road alignment is more the outcome of community politics than input from ACAP (Binod Gurung, ACAP official, Jomsom, personal communication, November 29, 2009).

This *ad hoc* type of road construction will inevitably lead to more landslides, which in turn can remove or destroy buildings, affect valuable productive agricultural land, and lead to the loss of livestock and human lives. This chain of events illustrates the coupled nature of road construction whereby an environmental impact can affect social systems in both economic and cultural spheres.

3.4.3 Local Spur Roads

Once the Kali Ghandaki Road was finished, it opened up the possibility for other spur roads to be built.

3.4.3.1 Tatopani

Tatopani is on the new Kali Ghandaki Highway 1 km upstream from where the Annapurna Circuit Trail (ACT) leaves the road and the Kali Ghandaki River Valley at Berrubowa (Fig. 3.5). Tatopani used to be a favorite stop for Nepalis and trekkers because of its geo-thermal hot springs. Upon reaching Tatopani trekkers looked forward to a soak in the naturally hot water. Guesthouses sprung up around the hot springs to provide trekkers with a place to stay and eat. The hot springs were also a popular spot for Nepalis and religious pilgrims to stop on their way up the valley to the sacred site at Muktinath. The new Kali Ghandaki Highway (Fig. 3.5) was built on top of the old trail, so the hot springs are now immediately adjacent to the road. Many of the trees that used to shade the area were cut down for the road right-of-way, changing what was an idyllic little oasis into a hot, dusty, noisy roadside attraction. The hot springs are leased on a yearly basis to concessionaires who bid for the right to manage them. The concessionaire who had been running the business for 2 years indicated that the number of both trekkers and Nepalis visiting this site had dropped significantly since the road was completed (K.C. Chhetri, lodge owner, Tatopani, personal communication, December 1, 2009). Another lodge owner commented that his business was also down because of the road. To compensate for the lost income he opened a roadside restaurant. Location played in his favor because buses and jeeps full of passengers stopped at his restaurant for rest and food after the long, bumpy 8–10 h journey down from Muktinath and Jomsom. Now, due to the flow of traffic on the road he is earning almost enough from the restaurant to make up for the losses at his guesthouse.

3.4.3.2 Berrubowa

Berrubowa (see Fig. 3.5) is where the popular Poon Hill portion of the ACT intersects the Kali Ghandaki Highway 1 km downstream (south) from Tatopani. This was the furthest extent of the preexisting road from Beni Bazar and the starting

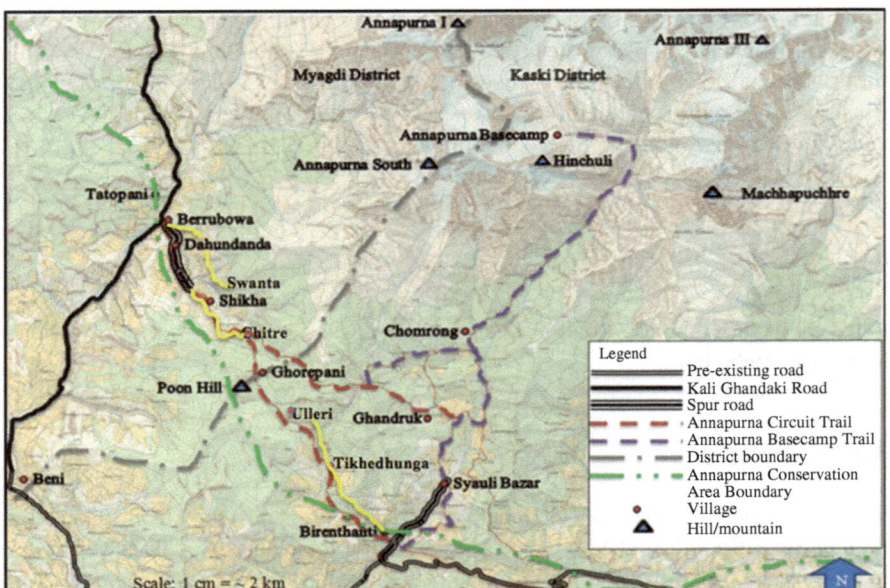

Fig. 3.5 Map of Annapurna Conservation Area in Myagdi and Kaski Districts with roads, spur roads, and trails. The *yellow lines* represent new local spur road building since 2009. (Adapted from Nepal-Kartenwerk der Arbeitsgemeinschaft für vergleichende Hochgebrifsforschug Nr. 9, 1993)

point of the new section of Kali Ghandaki Highway (see Fig. 3.5). Supplies were routinely transferred at Berrubowa from truck to storage houses or to mules for further distribution. Now that the Kali Ghandaki Highway is complete and construction of the spur road has started, goods can be taken much further by vehicle. Consequently, shop owners in Berrubowa are experiencing a decrease in business. The spur road is still under construction. At the point where it currently ends goods are transferred to mules, which transport the goods up the trail toward Ghorepani and Poon Hill. The spur road follows the old trail in places and veers off it in other places, until it reaches the top of the first ridge at Dahundanda (see Fig. 3.5). During the initial research period in the spring of 2009, the senior author hiked this section of the trail from Ghorepani to Berrubowa and then walked on the new Kali Ghandaki Highway to Muktinath. At that time, there were no signs of this new spur road. However, over the 6 months between research periods (April–May 2009 and October–November 2009), the spur road was begun, which was 10 km up the trail by the end of November. There were numerous landslide scars all along this section of new road and mule trains, trekkers, porters, and villagers had to jockey for position amid the loose rock and dust every time a jeep came along.

3.4.3.3 Dahundanda

Dahundanda (see Fig. 3.5) was an excavation site when the senior author visited it. The side of the mountain had been cut away to make way for the road. Walking up the road one could see a portion of the old trail buried under cut material that was dumped over the side of the hill. While there were not a lot of guesthouses in Dahundanda it was obvious from the debris, the excavation scar, and dust that not many people wanted to stay there. The owners of the two guesthouses interviewed reported a definite decline in trekker business, but they were pleased to have a road to aid in getting supplies in and out rather than having to carry them.

3.4.3.4 Shikha

The trail to Shikha (see Fig. 3.5) is a steep, steady, 4–5 h uphill hike from Dahundanda. The spur road extends several kilometres past Dahundanda and then fades into the traditional trail. Locals told the senior author that work had stopped because it was fall harvest season and as soon as they were done harvesting, they would continue building the road through the winter. They estimated it would reach Shikha by spring and possibly another 10 km past to Chitre. One of the lodge owners interviewed in Shikha expressed concern that as the road approached they would see fewer and fewer trekkers. She had already noticed a decline since the road had made it to Dahundanda. Local rumour has it that the road will eventually go all the way to Ghorepani (Lakshmi Bista, lodge owner, Shikha, personal communication, December 1, 2009).

Update 2014–2015

A recent field trip back to the area of the Kali Ghandaki Highway and Poon Hill (June 2014) confirmed that spur road building was expanding. Yellow lines in Fig. 3.5 represent the new road building. In addition to re-travelling the road from Tatopani through Jomsom to Kagbeni and Muktinath, this field trip continued north from Kagbeni to Lo Manthang in Upper Mustang (see Fig. 3.1).

Tatopani to Sikha

The spur road from Berrubowa where the traditional trail meets the Kali Ghandaki Highway, which had reached Shikha in 2009, has continued up the valley with sections both on and off the traditional trail and now reaches the village of Chitre (see Fig. 3.5). This road has been constructed by the local villagers and financed through the local VDCs. According to an ACAP official in Shikha the road may continue further, possibly all the way to Ghorepani where the popular Poon Hill trekking destination is located (Ram Gurung, Shikha ACAP officer, personal communication

June 10, 2014). He indicated that it was up to either the local villagers or the government, not ACAP, as to whether the road would go to Ghorepani. This issue is contentious because the lodge owners in Ghorepani do not want the road, as they are afraid it will decrease tourism business as it has in the areas where the road is currently. There is an alternate plan to route the road around Ghorepani and eventually have it reach Ulleri. One negative environmental impact of this new road that an ACAP official has noticed is the increase in the amount of trees that are being cut for firewood now that the road has provided easier access to the area (ibid). In addition, a new road could be seen across the river on the other side of the valley that reaches the village of Swanta (see Fig. 3.5) The resistance of lodge owners in Ghorepani to the road is not an isolated event. Lodge owners in Ghandruk (see Fig. 3.5) as well have halted the road at Syauli Bazar from reaching Ghandruk for the same reason; they fear the road will lead to a decline in the number of trekkers visiting their lodges.

3.4.3.5 Ghorepani

Ghorepani (see Fig. 3.5) sits on the top of a hill and has a commanding panoramic view of the Himalayas from Poon Hill, the highest point just above the village. This is an extremely popular trek and only requires a 3–5 day hike from the road at Birenthanti, the preferred starting point for the Poon Hill Trek. During the fall trekking season, it can be hard to find a bed in town if you do not arrive early in the afternoon. Ghorepani is situated perfectly to benefit from the road, as it is the obvious destination for tourists and trekkers. While it may lose a lot of its trekker-business as the roads get closer, it has the scenic pull factor to more than make up for that loss with tourists who will arrive by road if the road eventually reaches Ghorepani.

3.4.3.6 Syauli Bazar

Syauli Bazar (see Fig. 3.5) is near the end of the ACT. At this point, the majority of the trek is over with only a few hours of walking until one reaches the main road at Birenthanti. A spur road has slowly been expanding from the main road at Birenthanti along the trail toward Syauli Bazar. Now, trekkers can actually take a taxi along this section of the trail if they wish. During the time the senior author was there in 2009, the road was closed due to a flash flood that had destroyed a section of it. However, once that section of the road is rebuilt it will be extended further into the mountains, possibly toward Ghandruk and Ghorepani or maybe even the Annapurna Base Camp. Certainly, the people living in these villages would welcome a road.

3.4.3.7 Birenthanti

Birenthanti (see Fig. 3.5) is the village at the trailhead where the ACT meets the main road to Pokhara. It is the beginning or ending point of the ACT depending on which direction one walks. It is also the beginning point for various shorter treks along the

ACT such as the Poon Hill Trek and the Annapurna Base Camp Trek. Because of its location, Birenthanti has benefitted greatly from trekkers taking advantage of the lodging, food, and last minute essentials available in the multitude of shops that line the trail for the first kilometer. This is another example of ribbon development similar to what is taking place now at New Sangye (see Case Study #2) on the east side of the ACT. Even though this is primarily a trail, it is usable by taxis, as mentioned above. Taxi drivers go both directions on this spur road either to pick up tired trekkers finishing the ACT near Syauli Bazar, or to drop off trekkers who are just beginning the ACT at the same spot. In the future this spur road likely will reach Ghandruk and Ghorepani. Considering the spur road coming toward Ghorepani from Dahundanda, we expect that eventually these two spur roads will meet at Ghorepani, making it possible to arrive from either direction by road (see Fig. 3.5).

Update 2014–2015

Ghorepani to Birenthanti

Another road from Birenthanti has been extended up the Poon Hill Trail reaching Tikhedhunga and is continuing up the valley soon to reach Ulleri (see Fig. 3.5). The road track can be seen from Ulleri at a point across the valley where a bridge must be built to continue the road construction. Trekkers now walk on the road up to Tikhedhunga where the road leaves the trail and follows a different alignment. From Tikhedhunga to Ulleri trekkers can still use the traditional trail. In addition, the spur road that was heading from Birenthanti toward Ghandruk has now reached the village of Ghandruk.

Kagbeni to Lo Manthang

The Kagbeni to Lo Manthang road was not included in the 2009 research due to time constraints and the need for a special visa to enter Upper Mustang. During the 2014 re-visit to the area the senior author was invited by Brought Coburn of Colorado College as a road impact resource scholar to accompany a college student group trekking from Birenthanti to Lo Manthang. Brought was one of the first persons to enter Upper Mustang when it opened to western tourists in 1992 and has made numerous trips back into the area since. The following update is based on conversations the senior author had with him comparing the present road to conditions in the past (personal communication Brought Coburn, Colorado College June–July 2014) and interviews of local people conducted by the senior author.

Update 2014–2015

Numerous spur roads and short cuts were observed along the road to Lo Manthang. Due to the nature of the terrain lacking heavily forested areas it is easy to create short cuts through the numerous switchbacks on the official road. This shortens the time spent getting from one place to another and is relatively easy to do on a motor-cycle. The result is a plethora of tracks that diverge from the main road wherever it is convenient and expedient to do so (Figs. 3.6 and 3.7). In addition, local roads are being built with backhoes along previously existing trails to provide additional routes that are quicker and/or to provide access to areas that do not yet connect to the main road (Figs. 3.8 and 3.9). The by-pass road around Lo Manthang that was recommended and funded by the UNDP (Edwards et al. 2006) is currently not being used. Trucks are using the road near Lo Manthang's outer walls to enter the city; the same walls the UNDP was concerned would be adversely affected by vibration from trucks. This has promoted some lodge owners to consider taking action to reroute vehicles onto the UNDP recommended by-pass road (Sudan Gurung, lodge owner Lo Manthang, personal communication, June 25, 2014).

3.4.4 Summary

The coupled social and ecological system of road development can be seen in several examples of this road-building case study. In many cases the road has led to a decline in trekkers coming and staying in local trekking lodges because of the environmental impacts of noise and dust pollution as well as having to jockey for a space on the trail/road every time a vehicle passes. In one case however, a lodge owner has taken advantage of the road to open a roadside restaurant that is helping make up for the income he is losing at his lodge. Here we can see how location in proximity to a road can have multiple and conflicting socioeconomic impacts.

Fig. 3.6 Numerous tracks diverging from the main road near Samar, Upper Mustang (R. E. Beazley 2014)

Fig. 3.7 Divergent roads off the main road near Ghami, Upper Mustang (R. E. Beazley 2014)

Fig. 3.8 Trekkers on a new spur road near Syangboche, Upper Mustang. The main road, lighter in color, can be seen on the *right side* of the photo. (R. E. Beazley 2014)

Fig. 3.9 A horse and rider attempt to walk around a backhoe digging a new spur rode near Syangboche, Upper Mustang. (R. E. Beazley 2014)

Another important aspect that was mentioned numerous times is that in most communities only 10–20% of the inhabitants are lodge owners and for the overall population the roads provided many positive benefits such as ease of travel in and out of the area, better access to markets, health facilities, and educational opportunities, and lower transportation costs. Hence in one case the environmental impact of the road had a negative impact on trekkers and lodge owners' incomes but a positive impact on the majority of villagers in both socioeconomic and sociocultural terms.

Revisiting the Kali Ghandaki Highway and Poon Hill trek in 2014 showed a continuation of the spur road building observed in 2009 and new spur road construction in some other areas. Road building is continuing to be viewed as a threat to tourism by lodge owners and in several cases they have taken steps to curtail roads reaching their villages. Nonetheless, as was confirmed in 2009 the majority of inhabitants in most of the villages are not directly involved in the tourism business and the roads have brought many social benefits, such as better access to health facilities and education.

Along the Kagbeni-Lo Manthang road similar *ad hoc* spur road building is taking place leading to further landscape fragmentation and added dust and noise pollution. Lodge owners in Lo Manthang maintained that after the road arrived trekking tourism had declined. In addition many villagers who had traditionally raised horses and used them as means of travel, to transport goods, and for trekking groups have sold many of their horses as the road and vehicle transport has taken the place of the horse in their culture. As one interviewee put it, "With a motorcycle you only have to put gas in it when you drive it. With a horse you have to feed it every day whether you ride it everyday or not" (Tsering Wangyal, Lo Manthang resident and trekking guide, personal communication June 27, 2014). This same trekking guide remembers as child how she would make the long 10-day trek every year from Lo Manthang to spend the winter in Pokhara. Now locals can reach Pokhara from Lo Manthang by jeep in 1 long day.

In this case we can see the intricate connections between the environmental impact of a road having a direct impact on socioeconomic conditions in villages along the road and the sociocultural impact of eclipsing the horse culture, which has long been a part of life in Upper Mustang (Craig 2008). The socioeconomic impacts are differential. Those who are involved in tourism are feeling a negative impact on their income. However, after the arrival of the road from China and the arrival of cheap Chinese goods, it can be argued that the average Lo Manthang villager has experienced some positive socioeconomic benefits. For example, a quick price check on beer in shops in Lo Manthang revealed that Chinese beer was a third the price of Nepali beer. While the southern Kagbeni-Lo Manthang road may be causing a decrease in trekking tourists it has been credited with increasing domestic tourists who are arriving by jeep. Additionally it has reduced the cost of transportation as transport by vehicle is cheaper than transport by horse and the price of goods had also decreased substantially (Khadka 2014).

3.5   Case Study #2 Marsyangdi Highway

This case study examines some of the common environmental impacts of road building in mountainous areas including erosion, landslides and mass wasting, and ribbon development (Beazley 2013). As is discussed below these types of environmental processes can have serious impacts on socioeconomic factors.

3.5.1   Background

The Marsyangdi Highway follows the Marsyangdi River on the east side of the Annapurna Circuit Trail (ACT) and was built on top of the portions ACT on the west side of the Marsyangdi River (Fig. 3.10). The road has been built in stages over the years. It begins at Dumre where it connects with the Prithvi Highway that leads to Pokhara. The first section from Dumre to the Lamjung District Headquarters at Besisahar, a distance of 42.8 km, was begun in 1971 with the first vehicle reaching Besisahar in 1986; 15 years after the construction began (Full Bright Consultancy 2000). In 1992, the Nepal Army was given the task of building the section of road from the then current end of the road at Besisahar to the headquarters of Manang District at Chame, a distance of 65 km (Sharma 2011). The road was finally completed in 2012, 11 years after it was started (Sharma and Ghale 2012). During the

Fig. 3.10 Google Earth image of Marsyangdi Highway (*red line*), which follows the Mars
yangdi iver

2009 field research period the road had been extended from Besisahar to Sangye
and was progressing very slowly toward Chame due to several factors including the
extreme technical challenges as it approached the village of Tal, budget cuts, and the
political turmoil in the aftermath of the 10-year Peoples' War.

3.5.2   General Environmental Impacts

Numerous landslides (Figs. 3.11 and 3.12) were observed in 2009 along both the
completed section of the road to Sangye and the portion under construction past
Sangye. Villagers reported that the road is not passable during the monsoon due to
the frequent landslides. These landslides are not only a threat to personal safety, but
often take valuable agricultural land with them as well. One lodge owner noted that
the Marsyangdi Highway was dangerous due to landslides and that there had been
numerous injuries on the road including a jeep that rolled off the road plummeting
to the river below, killing all those inside (Ram Devkota, Bahundanda lodge owner,
personal communication, October 25, 2009). At the time of this fieldwork, environmental guidelines as well as best practice techniques for road construction were not
being enforced and hence severe erosion, landslides, and mass wasting will likely
continue to be a problem in the future.

Fig. 3.11 One of the many instances of landslides and erosion along the Marsyangdi Highway (R.E. Beazley 2009)

Fig. 3.12 Close up of another landslide along Marsyangdi Highway (R. E. Beazley 2009)

3.5.3   Sangye

Sangye is approximately 10 km upstream of Bahundanda and on the opposite side of the Marsyangdi River (see Fig. 3.10). It is strategically located at the end of the traditional ACT Bridge that crosses the river from the east side (Fig. 3.13) The bus stop is half a kilometer downstream of this bridge. Local villagers have taken advantage of the bus stop area to start businesses that provide services for arriving trekkers and Nepalis. This is a completely new section of Sangye, which was built very quickly after the road arrived in 2007 (Fig. 3.14). New Sangye consists of three new guesthouses, two new restaurants, a jeep taxi business, and a number of other new shops. Old Sangye has two guesthouses, in addition to a number of shops. The construction and alignment of buildings along this new strip development has been done in a very *ad hoc* manner, with apparently little planning or fore thought. The obvious environmental impact of the road is the landside area in between New and Old Sangye (Fig. 3.13). However, the rapid unplanned ribbon development of New Sangye (Fig. 3.14) will create environmental problems in the future unless careful measures are taken to deal with issues such as human waste from the new guesthouses and restaurants, vehicle fluids (leaking oil, brake fluid, and petrol) accumulation at the bus stop and jeep taxi business, and continuing ribbon development as businesses expand or new businesses are built.

3.5.3.1   Update 2014–2015

By 2013 the road reached Chame and motorcycles could reach Manang, and by 2015 jeeps were able to reach Manang (71 km) in 12 h. Only time will tell what will happen to New Sangye. New Sangye developed because the road had temporarily

Fig. 3.13 View of New and Old Sangye with landslide (R. E. Beazley 2009)

Fig. 3.14 New Sangye ribbon development. Old Sangye is just out of the far right side of the picture frame (R. E. Beazley 2009)

ended at Sangye due to the complications of building a very technical section upstream at Tal and budget cuts. In the future there may no longer need to be a bus stop at Sangye, in which case New Sangye may become a ghost town as travellers pass it by for more desirable destinations further up the road. In either event, New Sangye serves as a good example of the environmental impacts of ribbon development that is characteristic along slow moving *ad hoc* road construction projects.

Nonetheless benefits from the road are already being reported including lower transportation costs and lower prices for commodities. In addition the trip from Besisahar to Chame now only takes 6 h whereas before the road it could take 5 days or more. This has had direct health benefits, as before the road the only way to get to quality health care quickly was to charter a helicopter, a proposition not many villagers could afford (Lee 2015).

References

Adams, L. W. (1984). Small mammal use of an interstate highway median strip. *Journal of Applied Ecology, 21*, 175–178.

Amrhein, C., Strong, J. E., & Mosher, P. A. (1992). Effect of deicing salts on metal and organic matter mobilization in roadside soils. *Environmental Science & Technology, 26*, 703–709.

Aruga, K., Sessions, J., & Miyata, E. (2005). Forest road design with soil sediment evaluation using high-resolution DEM. *Journal of Forest Research, 10*, 471–479.

Beazley, R. E. (2013). *Impacts of expanding rural road networks on communities in the Annapurna Conservation Area, Nepal*. M.S. Thesis, Department of Natural Resources, Cornell University, Ithaca, NY

Brown, D. (2001). Characterizing the human imprint on landscapes for ecological assessment. In M. E. Jensen & P. S. Bourgeron (Eds.), *A guide for integrated ecological assessments* (pp. 404–415). New York: Springer.

Carson, B. (1985). *Erosion and sedimentation processes in the Nepalese Himalaya* (ICIMOD Occasional Paper No. 1). Kathmandu: ICIMOD. Retrieved from http://books.icimod.org/index.php/search/subject/15/45.

Cowie, I. D., & Warner, P. A. (1993). Alien plant species invasive in Kakadu National Park, tropical northern Australia. *Biological Conservation, 63*, 127–135.

Craig, S. (2008). *Horses like lightning: A story of passage through the Himalayas*. Boston: Wisdom Publications.

Dahal, R. K., Hasegawa, S., Nonomura, A., Yamanaka, M., Dhakal, S., & Paudyal, P. (2008). Predictive modeling of rainfall-induced landslide hazard in the lesser Himalaya of Nepal based on weights-of-evidence. *Geomorphology, 102*, 496–510.

Dahal, R. K., Hasegawa, S., Masuda, T., & Yamanaka, M. (2006). Roadside slope failure in Nepal during torrential rainfall and their mitigation. In H. Marui & M. Mikoš (Eds.), *Disaster mitigation of debris flows, slope failures and landslides*. Proceedings of the INTERPRAEVENT International Symposium Disaster Mitigation of Debris Flows, Slope Failures and Landslides held on September 25–27, 2006 in Niigata, Japan. Retrieved from http://www.interpraevent.at/palm-cms/upload_files/Publikationen/Tagungsbeitraege/2006_2_503.pdf.

Daming, H. E., Zhao, W., & Chen, L. (2004). The ecological changes in Manwan reservoir area and its causes. *Journal of Yunnan University, 26*(3), 220–226. Retrieved from http://www.lancang-mekong.org/Upload/upfile/20051226115012156.pdf.

Deoja, B. B. (1994). *Sustainable approaches to the construction of roads and other infrastructure in the Hindu Kush-Himalayas*. International Centre for Integrated Mountain Development. Kathmandu: ICIMOD. Retrieved from http://books.icimod.org/index. php/search/subject/25/60.

Edwards, P., Suwal, R., & Thapa, N. (2006). *Upper Mustang biodiversity conservation project: Final report of the terminal evaluation mission*. United Nations Human Development Programme (UNDP). Retrieved from http://erc.undp.org/**evaluation**admin/downloaddocument.html?docid=798.

Environmental Resources Management (ERM). (1996). *The significance of secondary effects from roads and road transport on nature conservation* (English Nature Research Report No. 178, (91)). Peterborough, ON: English Nature. Retrieved from http://publications.naturalengland.org.uk/file/115031.

Evink, G. L. (2002). *Interaction between roadways and wildlife ecology: A synthesis of highway practice* (NCHRP Synthesis 305). Washington, DC: Transportation Research Board—The National Academies. Retrieved from http://onlinepubs.trb.org/onlinepubs/nchrp/nchrp_syn_305.pdf.

Findlay, C. S., & Bourdages, J. (2000). Response time of wetland biodiversity to road construction on adjacent lands. *Conservation Biology, 14*(1), 86–94.

Forman, R. T., & Alexander, L. E. (1998). Roads and their major ecological effects. *Annual Review of Ecological Systems, 29*, 207–231.

Full Bright Consultancy. (2000). *Effectiveness of Investment in Dumre-Besisahar Road Final Report* submitted to is Majesty's Government National Planning Commission Secretariat Central Monitoring and Evaluation Division, Singh Durbar, Kathmandu.

Government of Nepal, Ministry of Federal Affairs and Local Development, Department of Local Infrastructure Development and Agricultural Roads (GoN/MoFAaLD/DoLIDAR). (2013). *District Transport Master Plan (DTMP) Mustang District*.

Hamiliton, L. S., & Bruijnzeel, L. A. (1997). Mountain watersheds-integrating water, soils, gravity, vegetation, and people. In B. Messerli & J. Ives (Eds.), *Mountains of the world: A global priority* (pp. 337–370). New York: Parthenon Publishing.

Harris, L. D., & Scheck, J. (1991). From implications to applications: The dispersal corridor principle applied to the conservation of biological diversity. In D. Saunders & R. Hobbs (Eds.), *Nature conservation 2: The role of corridors* (pp. 189–220). Surrey Beatty, Australia: Chipping Norton.

Hasegawa, S., Dahal, R., Yamanaka, M., Bhandary, N., Yatabe, R., & Inagaki, H. (2009). Causes of large-scale landslides in the lesser Himalaya of Central Nepal. *Environmental Geology, 57*, 1423–1434.

Hearn, G. J. (2002). Engineering geomorphology for road design in unstable mountainous areas: Lessons learnt after 25 years in Nepal. *Quarterly Journal of Engineering Geology and Hydrogeology, 35*, 143–154.

Hewitt, K. (1997). Risk and disasters in mountain lands. In B. Messerli & J. Ives (Eds.), *Mountains of the world: A global priority*. New York: Parthenon Publishing.

Ives, J. (2004). *Himalayan perceptions: Environmental change and the well-being of mountain peoples*. New York: Routledge.

Karan, P., & Iijima, S. (1985). Environmental stress in the Himalaya. *Geographical Review, 75*(1), 71–92.

Khadka, G. (2014, May 13). *Jomsom-Korala road boon for Mustang folks*. Kathmandu Post. Retrieved from http://www.ekantipur.com/2014/05/13/national/jomsom-korala-road-boon-for--mustang-folks/389492.html.

Khanal, N. (1996). *Assessment of natural hazards in Nepal*. ICIMOD's case study report (unpublished).

Landslide and Public Safety Project Team (LPSPT). (2001). *Forestry, landslides and public safety*. An Issue Paper prepared for the Oregon Board of Forestry by The Landslide and Public Safety Project Team. Retrieved from http://www.oregon.gov/ODF/privateforests/docs/LandslidesPublicSafety.pdf.

Laurance, W. F., Nascimento, H. E., Laurance, S. G., Andrade, A., Ewers, R. M., Harms, K., Luiza, R. C., & Ribeiro, J. (2007). Habitat fragmentation, variable edge effects, and the landscape-divergence hypothesis. *PloS One, 2*(10), e1017. Retrieved from www.plosone.org.

Lee, S. (2015, December 18–24) Taking the high road: Manang will not just survive, but prosper from its new road. *The Himalayan Times*. Retrieved from http://nepalitimes.com/article/Nepali-Times-Buzz/Manang-will-prosper-from-its-new-road,2749.

Lonsdale, W. M., & Lane, A. M. (1994). Tourist vehicles as vectors of weed seeds in Kakadu National Park, northern Australia. *Biological Conservation, 69*, 277–283.

Mader, H. J. (1984). Animal habitat isolation by roads and agricultural fields. *Biological Conservation, 29*, 81–96.

Merz, J., Dangol, P. M., Dhakal, M. P., Dongol, B. S., Nakarmi, G., & Weingartner, R. (2006). Road construction impacts on stream suspended sediment loads in nested catchment system in Nepal. *Land Degradation and Development, 17*, 343–351.

Merz, J., Nakarmi, G., Shrestha, S. K., Dahal, B. M., Dangol, P. M., Dhakal, M. R., Dongol, B. S., Sharma, S., Shah, R. B., & Weingartner, R. (2003). Water: A scarce resource in rural watersheds of Nepal's Middle Mountains. *Mountain Research and Development, 23*(1), 41–49.

Munguira, M. L., & Thomas, J. A. (1992). Use of road verges roads by butterfly and burnet populations, and the effect roads on adult dispersal and mortality. *Journal of Applied Ecology, 29*, 316–329.

National Trust for Nature Conservation (NTNC). (2008). *Sustainable development plan of Mustang (2008–2013)*. Kathmandu, Nepal: NTNC/GoN/UNEP. Retrieved from http://www.rrcap.unep.org/nsds/uploadedfiles/file/sa/np/mnmt/document/sd_masterplan_Mustang.pdf.

Noss, R. F. (1995). *The ecological effects of roads or the road to destruction*. Unpublished report. Wildlife CPR. Retrieved from http://www.wildlandscpr.org/node/41/print.

Oxley, D., Fenton, M., & Carmody, G. (1974). The effects of roads on populations of small mammals. *Journal of Applied Ecology, 11*, 51–59. Retrieved from http://www.jstor.org/stable/pdf-plus/2402004.pdf.

Petley, D., Hearn, G., Hart, A., Rosser, N., Dunning, S., Oven, K., & Mitchell, W. (2007). Trends in landslide occurrence in Nepal. *Natural Hazards, 43*, 23–44.

Reijnen, R., Foppen, R., Ter Braak, C., & Thissen, J. (1995). The effects of car traffic on breeding bird populations in woodland. III. Reduction of density in relation to the proximity of main roads. *Journal of Applied Ecology, 36*, 187–202.

Salick, J., Yongping, Y., & Amend, A. (2005). Tibetan land use and change near Khawa Karpo, eastern Himalayas. *Economic Botany, 59*(4), 312–325.

Schaffner, U. (1987) *Road construction in the Nepal Himalaya: The experience from the Lamosangu-Jiri Road project* (ICIMOD Occasional paper No. 8). Kathmandu: ICIMOD. Retrieved from http://himaldoc.icimod.org/record/7750.

Scheinder, R. (2010). Integrated, watershed-based management for sustainable water resources. *Frontiers of Earth Science in China, 4*(1), 117–125.

Shaffer, M. L., & Stein, B. A. (2000). *Precious heritage: The status of biodiversity in the United States*. New York: Oxford University Press.

Sharma, L. P., & Ghale, P. K. (2012 December 31). *Chame Road, a harbinger of development.* Kathmandu Post. Retrieved from http://www.ekantipur.com/the-kathmandu-post/2012/12/31/development/chame-road-a-harbinger-of-development/243544.html.

Sharma, L. P. (2011, November 21). *19 years on, Chame-Besisahar road construction still ongoing.* Kathmandu Post. Retrieved from http://www.ekantipur.com/the-kathmandu-post/2011/11/21/development/19-years-on-chame-besisahar-road-construction-still-ongoing/228472.html.

Sidle, R., Ziegler, A., Negishi, J., Nik, A., Siew, R., & Turkelboom, F. (2006). Erosion processes in steep terrain—Truths, myths, and uncertainties related to forest management in Southeast Asia. *Forest Ecology and Management, 224*(1–2), 199–225.

Smith, D. S., & Hellmund, P. (Eds.). (1993). *Greenway case studies: Ecology of. Greenways—Design and function of linear conservation areas*. Minneapolis: University of Minnesota Press.

Tsunokawa, K., & Hoban C. (Eds.) (1997). *Roads and the environment: A handbook* (World Bank Technical Paper No 376). Washington, DC: World Bank. Retrieved on from http://siteresources.worldbank.org/INTTRANSPORT/Resources/336291-1107880869673/twu-31.pdf.

Upreti, B. N. (2006). *The nexus between natural disasters and development: Key policy issues in meeting the millennium development goals and poverty alleviation* (Economic Policy Papers, Policy Paper 27). Government of Nepal/Ministry of Finance Singha Durbar, Kathmandu, and the Asian Development Bank. Retrieved from http://www.mof.gov.np/economic_policy/pdf/Natural_Disasters.pdf.

Upreti, B. N., & Dhital, M. R. (1996). *Landslide studies and Management in Nepal*. Kathmandu: ICIMOD. Retrieved from http://books.icimod.org/demo/index.php/search/type/0/136.

Vos, C., & Chardon, J. (1998). Effects of habitat fragmentation and road density on the distribution pattern of the moor frog *Rana arvalis. Journal of Applied Ecology, 35*, 44–56.

Wang, J., Cui, B. S., Liu, S., Dong, S., Wei, G., & Liu, J. (2007). Effects of road networks on ecosystem service value in the longitudinal range-gorge region. *Chinese Science Bulletin, 52*(Suppl. II), 180–191.

Wei, F., Liu, S., Degloria, S., Dong, S., & Beazley, R. (2010). Characterizing the "fragmentation–barrier" effect of road networks on landscape connectivity: A case study in Xishuangbanna, Southwest China. *Landscape and Urban Planning, 95*, 122–129.

White, P. A., & Ernst, M. (2003). *Second nature: Improving transportation without putting nature second*. Surface Transportation Policy Project, Defenders of Wildlife. Retrieved from http://repositories.cdib.org/jmie/roadeco/white2003a.

Wilcox, B. A., & Murphy, D. D. (1985). Conservation strategy: The effects of fragmentation on extinction. *American Naturalist, 125*, 879–887.

World Bank (WB). (2006). *Infrastructure: Lessons from the last two decades of World Bank Engagement* (Discussion Paper). Retrieved from http://www-wds.worldbank.org.

World Bank (WB). (2005). *Nepal North-South Transport Corridor Options*. International Development Association, Assistance Strategy Note. Retrieved from http://siteresources.worldbank.org/INTSARREGTOPTRANSPORT/34004316-1111699655514/20690624/NorthSouthCorridor03105.pdf.

Chapter 4
Socioeconomic Impacts of Roads

Abstract In theory, the economic benefit from road development accrues from the reduction in costs of both transport and travel. This allows several processes to take place that can have a positive impact on reducing poverty. Many factors can effect how the economic benefits of roads accrue to local populations and while there have been numerous studies to try to better understand this process many of these studies are flawed in different ways and do not lend themselves well to comparison. In Nepal results of some of these studies show varied and inconclusive results. One major study over a 20-year period showed very little concrete benefit from new roads. Other studies have shown that while new roads can benefit local populations they also can have unintended consequences, which may in the end negate some of the benefits. Finally, while roads are often touted as a solution to poverty alleviation the poorest of the poor usually benefit the least. Many of these effects can be seen along the Kali Ghandaki Highway in Mustang Districts where the highway has had impacts on diverse aspects of local economies including local tourism along the Annapurna Circuit Trail, religious pilgrimage practices at Muktinath, and changing land use and land values.

Keywords Socioeconomic Impacts • Connectivity • Markets • Poverty Alleviation • Transportation • Land Value • Differential Benefits • Kali Ghandaki Highway • Muktinath • Pilgrimage • Trekking • Tourism

4.1 Overview of Socioeconomic Impacts

In theory, the economic benefit from road development accrues from the reduction in costs of both transport and travel. This allows several processes to take place that can have a positive impact on reducing poverty. From an agricultural perspective, a road provides access to new markets for locally produced products in addition to access to new technologies, which may help maximize local production. Roads increase the volume and speed that goods can be transported, reducing the cost of transportation and the time to market. This allows farmers to grow perishable cash

© The Author(s) 2017
R.E. Beazley, J.P. Lassoie, *Himalayan Mobilities*, SpringerBriefs
in Environmental Science, DOI 10.1007/978-3-319-55757-1_4

crops as it reduces transportation costs and the risk of spoilage associated with moving their crops to a distant market, as well as the price they must pay for incoming goods and products (Richards 1984).

However, more than just a road is necessary for these advantages to accrue. This chapter considers what factors are necessary to optimize the socioeconomic potential of roads. It also investigates the question of who benefits from roads and how those benefits are distributed. Finally, it considers the unintended and unanticipated outcomes of roads by analysing previous case studies in Nepal in combination with our empirical research on the Kali Ghandaki and Marsyangdi Highways.

4.1.1 Poverty Reduction

Another important benefit from reduced travel and transport costs is enhanced ability to travel for employment both locally and to urban areas, which can then lead to poverty reduction. This enhanced ability to travel, in addition to the direct link to employment opportunities, is also beneficial in terms of building social capital and providing access to health and education facilities, which can also reduce poverty (Richards 1984).

Poverty reduction is often measured over a broad area without specific reference to what members in a poor rural area are benefitted. Poverty reduction does not necessarily mean that the poorest members of a community are the beneficiaries. For example, roads have been known to change land values dramatically creating both increased land consolidations and landlessness (Howe 1984). Increased accessibility may lead to an influx of cheap outside labor and consequent local unemployment. The use of vehicles may mean loss of employment for those who provided that transport before the road either through the use of animals or human porters. Some studies indicate that rural roads can result in greater inequality between poor and rich (Hettige 2006; Cook et al. 2005). Each situation has a unique set of circumstances resulting in winners and losers, making predictions about road programs often speculative at best. Many studies have been done on this subject because one of the main strategies for poverty alleviation in developing countries is the development of infrastructure, and roads are one of its basic building blocks. Without roads it would be very difficult to construct other types of infrastructure, such as schools, hospitals, telecommunications, and hydroelectric projects. In addition, roads provide access to other areas with road networks that facilitate people's movement. This in turn is thought to provide a number of social benefits, such as access to jobs and markets, enhanced social capital in terms of ease of contact with friends, family, and business associates, access to health and education facilities, and an easier way to travel and transport goods.

A review of donor institutions' reports from different developing countries highlights the importance of rural roads. "Inadequate rural transportation infrastructure and lack of mobility pose important constraints on rural development in Sub-Saharan Africa. Poverty assessments from Sub-Saharan Africa stress the close link between isolation and rural poverty"—World Bank (Calvo 1998: 7). "A key component of the Chinese government's poverty reduction initiative is to link the rural poor to major economic centers. Enabling poor people to benefit from greater mobility would increase their employment opportunities, open up trading and marketing of farm products, and improve access to health, education, and other social services."— Asian Development Bank (ADB 2006: 1).

Considering the number and scope of rural road projects in developing countries it is surprising that there is little agreement on the actual benefits; as Van de Walle (2008: 1) of the World Bank observes:

> In recent years, rural roads have been extensively championed as poverty alleviation instruments by the World Bank and donor institutions. It is argued that rural roads are key to raising living standards in poor rural areas (for example see Gannon and Liu 1997). By reducing transport costs, roads are expected to generate market activity, affect input and output prices, and foster economic linkages that enhance agricultural production, alter land use, crop intensity and other production decisions, stimulate off-farm diversification and other income-earning opportunities, and encourage migration. Claims have also been made that by facilitating access to social service facilities, better roads enhance social outcomes. Yet despite a general consensus on the importance of rural roads for development and living standards, there is surprisingly little hard evidence on the size and nature of their benefits, or their distributional impacts. Indeed, there have been relatively few rigorous and credible impact evaluations of rural roads.

Some common criticisms of rural road impact assessments are (Van de Walle 2008):

- Failure to follow projects for a sufficient time to document the full effects,
- Lack of proper comparison groups, and
- Not taking into account unobserved agents influencing the project in relation to the observed results.

4.1.2   Impact Assessments

The very nature of road projects makes impact assessment complicated leading to a variety of different methods and formulas being used during assessment. Consequently, it is very difficult to compare studies when different techniques are used for evaluation. While it is useful to identify some of these complicating factors, one of the overarching problems in assessment is that impacts can be short, intermediate, and long-term. Some impacts will be noticeable immediately while others may take a long time to appear making accurate impact assessment challenging. A road by itself does not impart benefits directly, but rather through the connections it makes with other complementary factors such as schools, vehicles for transport of goods, and demand for local goods. Due to the large number of factors involved in determining the benefits and the beneficiaries from road programs, assessment studies face the challenge of taking all of these derived and conditional benefits into account. Road projects are capital intensive and are planned to take advantage of certain aspects of their placement or for a specific purposes. Changes in the area once the road is built may not necessarily be the result of the road, or the results may be biased because of a pre-existing condition in the project area. Impacts from road projects can also vary greatly across different groups within an area and proper assessment most take into account all these different groups (Van de Walle 2008).

These are a few of factors that make accurate road project assessment complicated and conflicting. Howe (1984: 50) points out that it is very difficult to assess the validity of these project studies without "...highly sophisticated experimental controls and equally sophisticated analysis." After reviewing numerous road assessment reports he concluded that: "Rarely has the design of any impact study approached these 'necessary' scientific standards. Thus it must be expected that results will usually be tentative and contradictory and that only broad conclusions will be possible." (Howe 1984: 53).

It is interesting to note that the comments made in 1984 by Howe about the lack of impact studies with 'necessary' scientific standards, is very similar to the comment made 24 years later by Van de Walle (2008: 1) when he said: "Indeed, there have been relatively few rigorous and credible impact evaluations of rural roads." This again may be an indication of the complexity of interrelated factors involved in road projects and the inherent difficulty in assessing the impact of rural road projects. Van de Walle (2008: 30) summarized the essential elements of this complexity by saying:

> The benefits of rural roads are indirect and conditional on interaction with the geographic, community and household characteristics of their location. Road locations are typically determined by those same characteristics confounding inferences based on comparisons of places with roads versus without them. Additionally, impacts may be distributional, felt across multiple outcomes and take a long time to emerge. These features of rural roads have implications for evaluation design and data collection.

Even when rigorous methods have been used for project evaluation there still appears to be ambiguity on overall outcomes of rural road impacts.

> In recent years a number of studies have assessed the impacts of rural roads rigorously using impact evaluation methods that expressly deal with selection. These studies show mixed results, some finding substantial impacts and others more muted impacts. They have examined disparate outcome variables, in diverse circumstances, using various techniques—some of which can be questioned. Sources of ambiguity in impacts can also be expected due to heterogeneity. It thus remains difficult to draw definitive conclusions concerning the impacts of rural roads. (Van de Walle 2008: 26)

Therefore, it must be recognized that each road project is extremely context specific and while it may be informative to review previous rural road project assessments, drawing conclusions that can be applied to other countries or even other areas in the same country have many pitfalls. Nevertheless, this body of impact assessments forms the existing hard data on the socioeconomic rural road project outcomes. Consequently, it is useful to review it, but with the above mentioned limitations in mind.

4.2   Socioeconomic Impacts in Nepal

Several particularly interesting rural roads studies have been done in Nepal that looked at road development in the same area over a 20-year period. In 1971 an *ex-ante* study of the Sonauli-Pojhara Highway near Pokhara was done to "...determine the potential of the individual farmer to respond to new technology, government programmes and changes in absolute and relative prices" (Schroeder 1971: 4). After doing an in depth econometric analysis they concluded that the new road was an enabling mechanism that could create economic benefits in the Pokhara Valley in the future. The road provided the means by which the area had "...the opportunity to progress rapidly enough so that production and regional income increase faster than population growth" (Schroeder 1971: 6).

In 1973, the Overseas Development Group of the University of East Anglia undertook an extensive study of Nepal's road system, titled the Nepal Roads Research Project (Blaikie et al. 1976). This detailed 3 year study's objective was "...to investigate the social and economic effect of road provision, with particular reference to inequality, in the west central region of Nepal." (Blaikie et al. 1976: 1.1) The study included the section of road studied by Schroeder (1971) mentioned above as part of the East-West Highway (Mahendra Rajmarga) development. In addition, two other sections of road were studied that would connect the north hill regions, one (the Prithivi Rajmarga) with Kathmandu, and the other (the Siddartha Rajamarga) with the plains and India. These last two were included by the request of His Majesty's Government of Nepal to determine if: "...the building of roads linking, different and unequal, regions would have the effect of reducing inequalities between those regions, and also, through their more general economic and social effects, between individuals and groups within each regions' as stated in the fourth Nepalese Five Year Plan" (Blaikie et al. 1976: 1.1–1.2).

At the time of the study, Nepal was facing an escalating poverty trend exacerbated by a rapidly growing population and a 10-year decline in food grain yields. This was highlighted by the 1974 U.N. report, which said: "Nepal is poor and is daily becoming poorer" (ARTEP 1974 cited in Blaikie et al. 1976: 5). The hope of the Nepalese government was that road construction would be a solution to this problem by evening out some of the regional inequalities and stimulating agricultural and industrial growth. In the summary of the findings of the report indicated that:

> The building of three main roads in West-Central Nepal over the last ten years has had very little effect on the crucial pre-requisites for any significant development in the region, namely increasing agricultural production and industry. There have been some changes in commerce and trade, but in part to the growth of population and the bureaucracy. No evidence has been found to support the optimistic prognosis of the effect of roads in the fourth Nepalese Five Year Plan. (Blaikie et al. 1977: 131)

The reason for this failure of road development to deliver the results forecasted had to do with several factors including Nepal's trade treaties with India, which favored India at the expense of Nepalese forms of production, the reliance of the government on foreign aid and the corruption inherent in that system in Nepal, and the inequality of the class structure. This study then became the basis of the book *Nepal In Crisis* (1980), which predicted that Nepal would become more poverty stricken, as its economy slowly continued an inevitable decline and roads would not alleviate this. The mitigating factors were so strong that the authors stated that: "The roads would have effects but these would essentially serve to deepen dependency and underdevelopment rather than alleviate it" (Blaikie et al. 2002: 1256).

Further analysis showed that no matter how good the roads were, unless there is complimentary development and support by the government or private investment to increase production, road benefits would not accrue. In Nepal the government had not implemented sufficient programs to aid poor farmers and the level of poverty was so high farmers were not willing to take on the risk of innovation without some form of support (Leinbach 1995). The situation was complicated by Nepal's proximity to India and India's preferential trade agreements, resulting in more imports from India, which stunted Nepalese production and industrial growth.

Blaikie et al. (2002) re-examined this situation in 1998, more than 20 years after their original study. Although the prediction of increasing poverty made in *Nepal In Crisis* (Blaikie et al. 1980) was not evident, there still had been very little change: "...hardly any significant development of commercialized agriculture, little investment in modern inputs, aggressive forcing down the cost of hired labor, or appropriation of poor peasants' land by an expansive class of capitalist farmers" (Blaikie et al. 2002: 1267). One explanation for the lack of development in agriculture was that it was financially better to have family members migrate, either domestically or internationally, for work than it was to enter the market as a producer because it was less risky and pays better. This trend in migration had been increasing since the first study, almost tripling over 20 years. However, the authors failed to mention that roads aided in the migration trend, which helped preclude the deepening slide into poverty that the original study predicted. Another factor influencing the reluctance to invest in agriculture was that many farms in Nepal are so small that they do not produce enough surpluses to establish a base for major investment.

A study done in Nepal in the 1990s by the International Centre for Integrated Mountain Development (ICIMOD) (Paudyal 1998) was designed to look at several aspects of rural roads to determine how they could be used more effectively as a means of poverty alleviation and sustainable development. It examined not only the effects of rural roads on local villages, but also the conditions that affected rural road construction, functioning, and maintenance. Consequently, policy, institutional responsibilities, and production linkages were also analysed. Four rural road projects from different areas of Nepal were picked to reflect different levels of connectivity to markets and different construction techniques. Analyses of the case studies were done at three levels: policy, program, and project.

At the policy level the report pointed out the need to keep in mind the institutional linkages in the rural road system when planning rural road policies and studies. At the time of the study, the implementing institutions for rural roads were the District Development Committee (DDC) and Village Development Committee (VDC). However, the operational plans for road projects failed to take this into account. Therefore, many of their recommendations on how to carry out the project were unrealistic at the district and village levels. At the program level, while the local institutions (DDC) had all the elements contained in the overall road development plan, at the project level many of the steps involved in the project were ignored because the DDC and VDCs are autonomous.

Nepal's Department of Roads (DoR) has a five-stage protocol for proper road construction, but at the local project level, it is rarely enforced. One example given is that phased construction, which allows for natural stabilization of the road over time and provides environmental protection, is often ignored because construction is based on budget allocations. Consequently, phases are not interlinked in the proper way resulting in poorly constructed roads that are hard to maintain. This is compounded by the fact that the maintenance policy for rural roads is not clear. Without having an overseeing regulator agency that can enforce proper construction techniques at the local level many of these poorly built roads lead to erosion and landslides, and have other negative impacts on the environment. Often roads are built in the fastest, least expensive way so vehicles can begin using them as soon as possible.

At the project level, benefits were reported, such as decreases in transportation costs, but apart from kerosene and salt, the price of staple goods did not decrease. Availability of goods in the area did increase, but the base price of these goods did not fall below pre-road prices, and in fact increased over time. The savings in transport cost were absorbed by transportation owners and retailers, making them the main beneficiaries, not the consumers who had provided both free land and labor for building the road. In terms of positive social impacts, both education and health benefits were noticed because better teachers were able to move to the area and the amount of time needed to reach the hospital was reduced. An additional overall benefit of people feeling less isolated and more connected to the outside world and its services was also reported.

Recommendations included in the report pointed to several important issues for rural roads in Nepal. First, the concept of rural roads must include more than just roads that vehicles can use. It should also include interconnected trails that are used to reach the nearest motorable roads and how that road connects to the rest of the nation's road network. Without taking this overall pattern into account the full potential of rural road construction and expanding networks cannot be actualized.

Secondly, the institutional connections for planning, construction, implementation, and maintenance of rural roads lack clarity and coordination. Local level institutions should submit their needs to the national level, which in turn can provide support and the necessary supervision to ensure the project progresses efficiently using both sound technical and environmentally positive techniques. This should also involve employment of the local population in the construction of the road, which would accrue additional benefit to the local area. Creating an institution at the district level accountable to the local level to oversee the coordination of the various phases of the project and the continued maintenance and functionality of the roads after construction was completed would facilitate the connection between the local and national levels.

Thirdly, local government needs to ensure that the road will increase local production by coordinating the various actors involved in production, transportation, marketing, and storage of local goods.

Lastly, several environmental and spatial considerations were deemed critical. The construction of rural roads would benefit from following the Low-cost, Environmentally-friendly, and Self-help (LES) approach found to be successful in other areas of Nepal. This would have positive benefits for the local population and help reduce the amount of negative environmental impact. LES is a comprehensive program that employs local people under supervision by engineers to construct a road using locally available materials and environmentally friendly techniques, such as constructing in phases, balancing cuts and fills, water runoff management, and bioengineering for slope stabilization. This approach gives communities a sense of ownership of the road that helps ensure proper maintenance after the road is complete. In terms of spatial considerations, setting up a facility at the road head that could be used for storage and distribution of goods and help centralize and organize other ensuing construction at the road head, would reduce the tendency for uncontrolled sprawl and ribbon development along the road (Paudyal 1998).

Various NGOs in Nepal are involved in rural road projects and reports on their projects provide some of more recent information about the country's road development.

The Swiss Agency for Development and Cooperation (SDC) has been working in Nepal for over 50 years. In 1999 it initiated the District Road Support Programme (DRSP) to focus on "… improving road access while benefiting the people who really need it. It aims to build the capacity of the participating district organizations to plan, design and implement the maintenance, rehabilitation and construction of district roads".[1] This program covered six districts in eastern Nepal. In 2009, an independent external review of the program was published (Strickland 2009). One of the main objectives of the review was to determine the economic and social impact of DRSP on the local communities. Four roads from three of the districts were chosen to be representative of a wide range of geographical, access, and transport characteristics. The overall conclusion was that the DSRP road project had a positive impact on local communities. This was observed in both economic and social empowerment. Economic growth was enhanced directly through employment of the locals during the road construction phase, and this was found to favor the poor. Other economic benefits were accrued in terms of reduced transportation cost, which led to lower prices of imported goods, additional employment options, and access to new markets outside the local area for selling farm produce. Social benefits included greater equity for marginalized groups and women, from empowerment through community activities facilitated by the road linkages, and general social benefits associated with better connectivity. The report also mentions that while the poor have benefitted from employment during the road construction phase, these benefits are short-term and long-term benefits tend to favor more economically advantaged groups due to their ability to invest in the new opportunities resulting from increased connectivity (Strickland 2009).

---

[1] http://www.swiss-cooperation.admin.ch/nepal/en/Home/District_Roads_Support_Programme_Phase_3.

A study by the Asian Development Bank (ADB) in 2007 (Kafle 2007), emphasized the importance of integrated road construction projects, such as the SDC project mentioned above, for maximizing benefits of rural road projects for poverty alleviation. ADB conducted a pilot project in three districts in eastern Nepal named Enhancing Poverty Reduction Impact of Road Project (EPRIRP). A key element was involving four NGOs with the DoR in organizing micro-finance, skills training, and income generating activities for poor and marginalized groups along the different road sections. The positive benefits that accrued to such groups have led ADB to recommend that NGOs be involved in future road projects aimed at poverty reduction (Kafle 2007).

Using data from Nepal, Jacoby (2000) developed a model based on farmland value in relation to distance to farm produce markets to predict the benefit of a hypothetical road project. His results showed that expanding rural road network connections from farm to markets would lead to a sizable benefit. However, while the poor could earn a substantial amount of this, the size of benefit would not be large enough to offset income inequalities in the area.

Malla and Griffin (2000, cited in Ives 2004) studied four communities with varying road accessibilities. The first village, Dhulikhel, was well connected by the main road to Kathmandu. The second village, Chhattrebanhjh, was connected to Dhulikhel by road. The last two villages had no road connections; Chaubas is a 6-h walk to the nearest road and Budhakani is more than a 1-day walk to a road. The study covered 102 households and included extensive data on socioeconomic conditions, livelihoods, industrial development, and overall change over a 10-year period. Their general conclusion was that road accessibility was the defining element in the differences between the socioeconomic conditions of the villages.

The most outstanding difference over the 10-year period was the large increase in commercialization and trade in the two villages connected by road. In addition, increased agricultural intensity was attributed to the availability of chemical fertilizers in both of the road-connected areas, with households reporting self-sufficiency that had none a decade before. With the ability to grow surplus food for sale, choice of crops changed to meet market demand. This trend was also noticed in livestock production, with meat and milk being favored. In contrast, farmers in the two villages not connected to a road grew food primarily for home consumption; although Bukhakani did supply a substantial number of goats for sale in Kathmandu. Dhulikhel, being the closest to Kathmandu, had diversified considerably by adding numerous new businesses (e.g., carpet making, hotels, brick making, transport, and furniture making) with an attendant growth in employment opportunities. An impact on education also was noted with most children now attending school. The other important observation was that the degree of road access had a major influence on the ability to derive income from off-farm employment, which increased overall household security for those with road access.

In contrast, a study by Bohle and Adhikari (1998) was done in two areas in west central Nepal specifically chosen for their relative inaccessibility to markets. Three to five villages were investigated in each area representative of the most remote villages. In one area, only 8% of the households had food surpluses, with 51% being self-sufficient from their own production less than 12 months and 41% for less than 6 months. In the other area, only 1% had food surpluses with 31% self-sufficient for 7–12 months and 68% for less than 6 months. Purchasing and bartering of food were the main ways people dealt with the food deficit. While none of these villages were directly connected by road, the authors were able to determine one of the key elements in livelihood survival was the ability to be mobile, both spatially and seasonally, to take advantage of income generating and trading opportunities between villages along the trails, and to access different ecological niches, which provided additional coping strategies. Nevertheless, indications from a caloric intake assessment showed that even in the best cases, villagers were only able to achieve 85% of the minimum food requirement established by the World Health Organization (Bohle and Adhikari 1998).

Seddon and Shrestha (2002) studied a village along the Bhimdhunga-Lamidanda Road in the central region of Nepal. Major benefits reported were decrease in transportation costs and travel times, with the increase in personal mobility considered the most important benefit. Several roadside effects were identified. Landowners along the road or immediately adjacent to it were the first to benefit due to the major increase in their land values. This translated into home improvements and an increase in consumer goods such as TVs and radios. Traders and vehicle owners were the next group to benefit, although large trucks for transportation tended to only be affordable by wealthy merchants from Kathmandu. Many new businesses along the road providing services for travellers were built, such as restaurants, shops, and small lodges. An almost complete loss of work for porters in the village came with the road, except during the monsoon season when the road was impassable. Before the road, porterage provided valuable work for many of the poorer inhabitants of the village. Agricultural production also changed since the road opened. With the availability of fertilizers and connection to different market demands in Kathmandu several changes in crop production occurred including growing high yield rice instead of the tradition rice variety, and more production of vegetables and fruits. Additionally, eating habits had changed from maize and millet, to the main stable of rice, now that cheap rice was more readily available.

Part of this success in increased agricultural production is due to the Small Farmers' Development Programme, which provided low interest credit to farmers. This production increase has been so successful that vegetable production has taken over fields that used to be sown with rice and wheat, in addition to some land being leased for vegetable production. Social benefits cited include increases in personal travel for family, social, and religious activities, and in literacy rates in road access areas versus non-road access areas, due to more primary schools being built along the road (Seddon and Shrestha 2002).

The importance of complimentary services to help stimulate the agricultural intensification benefit of rural roads has been highlighted in other mountainous areas of Asia, such as Hunza in the Karakoram (Kreutzman 1991).

A study by Brown (2003) used field-based surveys with GIS data to examine the effects of a road on socioeconomic issues and gender roles in eastern Nepal. The study was conducted in the Yarsha Khola watershed along the Lamosangu Jiri Road, which links the area with the capital, Kathmandu. It compared the area along the road with the adjacent area, which had no road access. Significant impacts were found as a result of the road when comparing the two areas. Specifically, farmers in the area with road access used more chemical fertilizers, had smaller land holdings, and relied more on off-farm income to support them.

4.3 Summary

As Howe states in *The Impact of Rural Roads on Poverty: A Review of the Literature* (Howe 1984: 79): "The over-ridding impression gained from reviewing available impact studies is the paucity of evidence on the effects of road investment programmes on rural incomes and income distribution." This statement points to one of the main difficulties in determining the impact of rural roads on socioeconomics. However, it is not only the lack of information; it is also the unreliability of the information available due to poor design of case studies examining this issue. In addition, the lack of consistency between different studies makes it hard to compare and contrast their results. However, a summary of the literature on socioeconomic impacts of rural road construction suffices to point out many of the same scenarios in other developing countries.

The literature on socioeconomic benefits from rural roads in Nepal is instrumental in pointing out the numerous contributory factors that must be present for rural roads to benefit an area economically. The pre-existing political situation in Nepal, in addition to its treaties with India on its southern border, has meant that many of the roads that have been built to help develop rural areas were not able to stimulate the economic growth necessary to bring major benefits to the area. This is complicated by the historical inability of the government to develop agricultural aid programs in rural areas that would provide the support necessary to allow farmers to take the risks involved in changing their mode of production to maximize the new opportunities for market access that the roads provide. In fact, the increased mobility brought about by roads has resulted in more migration in search of off-farm employment, as this has proven to be more profitable and less risky.

Recent trends in collaborative efforts with local communities to build roads have had differing results. In places where careful planning and supervision were involved due to NGO involvement in integrated pro-poor programs, results have been positive, providing employment for disadvantaged groups. Where this structural organization was lacking, the resulting *ad hoc* nature of local road building has often resulted in substandard roads that are hard to maintain and result in serious environmental degradation.

Some of the negative socioeconomic outcomes of roads when they do not have complimentary economic stimulus packages are stagnation, unequal distribution of benefits, and landlessness.

4.4   Case Study #1 the Kali Ghandaki Highway

This case study shows how a road can influence both land use and land value in five
different villages in Mustang District: Muktinath, Jharkot, Jhong, Chungkhar, and
Putak (Fig. 4.1) (Beazley 2013). However, it should be kept in mind that in Hinduism
and Buddhism, land uses and values are not just associated with economic value,
but with spiritual values as well. "The sacred is a genuine land use for Tibetans"
(Salick et al. 2005: 319).

4.4.1 Muktinath

The land use and land value changes occurring in Muktinath are because of the road,
which provides a cheap and efficient way for religious pilgrims to visit the Muktinath
temples. Since the Kali Ghandaki Highway and spur road to Muktinath was completed in 2008 the number of pilgrims has increased by three times (Shiva Ram Shuba,
Head Hindu priest Muktinath, personal communication, November 27, 2009). Now, a
special facility is needed for pilgrims to cook food according to their religious tradition and for lodging. In the Hindu religion, Brahmins have very particular ways in
which their food should be prepared and served that guesthouses in Muktinath are not
able to offer.

Fig. 4.1 Location of the villages of Muktinath, Jharkot, Jhong, Chungkhar, and Putak in the Jhong
Khola River Valley in Mustang District

In response to the increase in pilgrims and the need for appropriate lodging and food preparation facilities, the governments of India and Nepal (Fig. 4.2) are building a pilgrim's *dharamsala* with lodging, personal cooking space, and a meditation—worship room. The government of India offered to finance the building of the *dharamsala* if the community would donate a piece of land. The villagers formed a community land organization and chose a plot of barren community land near the beginning of town to be the site for the *dharamsala*. They also negotiated an agreement with the Indian government to manage the *dharamsala* and distribute the income among the community members (Krishna Shrestha, lodge owner, Muktinath, personal communication, November 28, 2009). In this way, the road drastically changed the use and value of a piece of community land that was barren and unused. Hence, it will not only produce community income but will also create religious merit for the community.

The construction of the *dharamsala* has led to another land use change. Construction has begun on a spur road, from the jeep stop/stand at the beginning of town to the *dharamsala*, and then on to the Muktinath temple complex on the hill (Fig. 4.3). This will not only make it easier for pilgrims to get from the *dharamsala* to the temples, but it will also reroute traffic around the town rather than directly through the town's center, which is the route that the motorcycle taxis presently use. We believe this will eventually lead to a network of spur roads in this area.

The other interesting land use/value change in Muktinath, stimulated by the road and the use of both motorcycle taxis and jeeps, is the construction of a ticket office with a parking area at the beginning of town. This also serves as a staging point for motorcycle taxis that take tourists and pilgrims who arrive by jeep up to the temples.

Fig. 4.2 The sign at the site of the new *dharamsala* in Muktinath (R. E. Beazley 2009)

Main part of town

Muktinath Temple

New road for pilgrims from the *dharamsala* to the Muktinath Temple

New road from the jeep stand to the pilgrims' *dharamsala*

Fig. 4.3 New spur roads for pilgrims from the jeep stand to the pilgrims' *dharamsala* and on to the Muktinath Temple complex (R. E. Beazley 2009)

4.4.1.1   Update 2014–2015

Several changes have occurred in Muktinath since 2009, which provide further evidence of continuing land use/value changes. The jeep stand that was on the main road leading into Muktinath has been moved to be nearer the pilgrims' *dharamsala* (Fig. 4.4). A new road cut has been made off the main road to funnel traffic to the jeep stand (Fig. 4.5). Another interesting change is that the motorcycle taxis that had taken religious pilgrims from the jeep stand to the bottom of the stairs leading to the Muktinath temples now have a new track that allows them to take pilgrims all the way to the top of the stairs. Hence, pilgrims no longer need to climb the stairs to reach the temples, a difficult task for the elderly and infirm. In addition, the 7-km Kagbeni to Muktinath road started upgrading to a black top road beginning in December 2014 and is scheduled to be finished by 2017 at a cost of Rs 208 million (Phalebus 2015).

4.4.2   *Jharkot and Jhong*

These two villages are located opposite each other in the Jhong Khola River Valley approximately 5 km downstream from Muktinath (see Fig. 4.1). They make a very interesting comparison because the road is complete on the Jharkot side of the valley whereas the Jhong side has no road, but it does have an alternate trail down the

Fig. 4.4 The new jeep/bus stand relocated off the main road, now near the pilgrims' *dharamsala* (the building in the background) (R. E. Beazley 2014)

Fig. 4.5 The new road construction to the new jeep/bus stand (R. E. Beazley 2014)

valley that trekkers are now starting to use to avoid walking on the road. There is a road planned to take the place of this trail and become part of a part of the Muktinath Valley Ring Road. In both villages, there is evidence that the road has influenced changes in land use and land value.

4.4.3 Jharkot

Muktinath is the first village trekkers reach after coming over the Thorang La Pass from the east side (see Fig. 4.1). Exhausted trekkers coming down from the pass cannot wait to put up their feet and have a cold beer while listening to reggae in the Bob Marley Guest House. Jharkot is another 15 min down the trail and has quite a few nice modest guesthouses, an old monastery, and a fortress. Jharkot has been a favorite of trekkers who do not want to deal with the hustle and bustle of Muktinath. Guesthouse owners in Jharkot all agreed that the number of trekkers has decreased since the road was completed. One of the main reasons is that trekkers who are tired of hiking after crossing the Thorang La Pass can now hire a jeep in Muktinath to take them down the road rather than walking all the way to Jomsom on the road. The other reason is that trekkers do not want to walk on a dusty, noisy road. The decrease in trekkers has led the lodge owners to seek additional livelihood options. Several of these options include collecting the caterpillar fungus *Cordyseps sinensis* known as *yartsagumba* in Nepal (Fig. 4.6), harvesting sea buckthorn (*Hippophae rhamnoide*) to make into juice, and expanding their fruit orchards.

The road has influenced their pursuit of these livelihoods in reciprocal effects and feedback couplings in that it has decreased their trekker business, but at the same time increased their ability to pursue livelihood options that they could not before the road (Fig. 4.7). In the past, it took 5 days by donkey to get produce to market, which meant most crops were only sold locally. With the road, farmers can now expand their present crops and grow other crops that were not economically viable before. Several guesthouse owners/farmers have made a land use change by expanding their fruit tree orchards now that they can use the road to move the fruit to market before it spoils. Other farmers can now more easily get supplies from bigger distribution centers such as Pokhara, in addition to getting their produce to market faster and cheaper. This will eventually lead to speciality-market mono-cropping based on demand, one of the more predictable outcomes of rural road construction in agricultural areas (Salick et al. 2005; Allan 1986). Another predictable impact of rural road construction is a change from using animal fertilizer to the use of cheap chemical fertilizers. The road provides a link to fertilizer distribution centers and lowers the cost of transportation, making fertilizer more affordable to farmers.

There are additional indications of land use changes being influenced by the road in Jharkot. In an attempt to attract more trekkers, villagers have erected signs describing some of the unique features of their village (Fig. 4.7). These did not exist during the first research period (April–May 2009). This change of land use, while minimal compared to expanding an apple orchard or building a *dharamsala,* indicates the

Fig. 4.6 Caterpillar fungus (*Cordyseps sinensis*) known in Nepal as *yartsagumba*, highly valued in Chinese medicine (R. E. Beazley 2009)

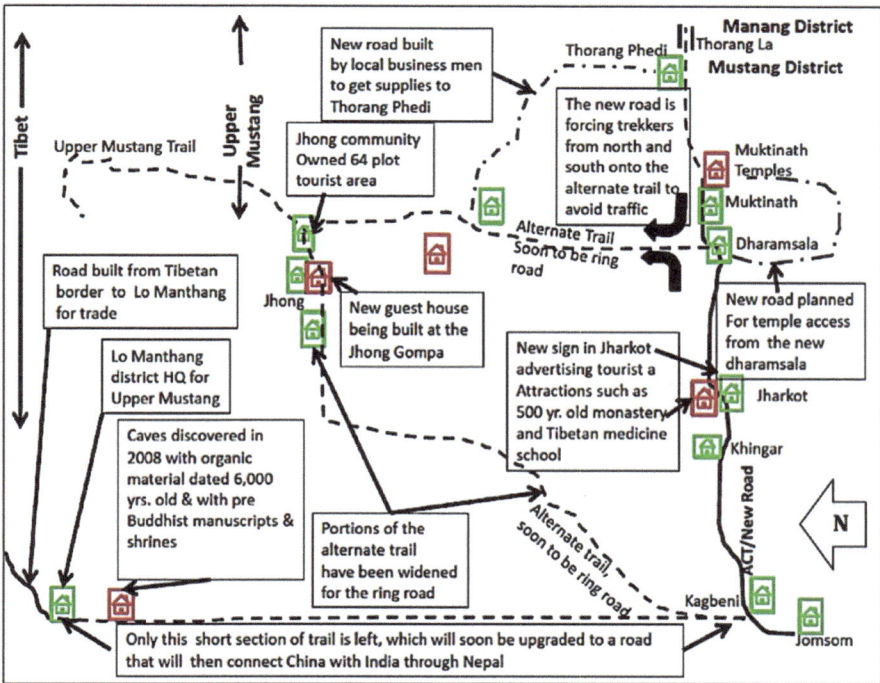

Fig. 4.7 Summary diagram of Case Study 1 (Beazley 2013)

impact the road had over the relatively short 6-month period between research trips. One of the places of interest mentioned on the signs is the 500-year-old Buddhist *gompa* (see Fig. 4.7) that tourists can visit with local guides who speak both English and French. Inside the *gompa* walls is a Traditional Tibetan Medicine School (see Fig. 4.7), another tourist attraction mentioned on the sign. There is a famous traditional Tibetan medicine doctor (*amchi*) who gives classes in Tibetan medicine and who will give visitors a diagnosis if they wish.

Another land use change is the renovation of an old abandoned fortress that was in the process of falling down. The village development committee decided to repair and expand it into a museum of local culture and a new community center. The road is aiding in this reconstruction by making building materials easier and cheaper to transport to the renovation site.

Even though all the lodge owners interviewed in 2009 said business was down, there was one new lodge under construction, an indication that some villagers think business will eventually pick up.

4.4.4 Jhong

Jhong (see Fig. 4.7) technically lies in Upper Mustang District, but it is so close to the border of Mustang District that the locals have persuaded the Annapurna Conservation Area Project (ACAP) to allow trekkers to take the 45-min hike across the valley to visit Jhong from either Jharkot or Muktinath. Curious trekkers, who still have energy after crossing the Thorang La Pass, sometimes cross the valley to visit Jhong on their way down the alternate trail (see Fig. 4.7) that leads to Kagbeni. The new Muktinath road on the Jharkot side of the valley is influencing more trekkers to visit Jhong and to use the alternate trail rather than walking on the road.

The villagers of Jhong realize their potential to benefit from more trekkers who choose to cross the valley to the alternate trail and stay off the road. They are so optimistic about the trekking business increasing that they have divided a piece of community property into 64 plots for guesthouses and tourist facilities (see Fig. 4.7). Now that the Muktinath Road is complete, Jhong has the perfect location to benefit from an increased number of trekkers, not only because the road entices trekkers to cross from the other side of the valley, but also because they are situated on one of the trekking trails that comes down from Upper Mustang (see Fig. 4.7). One of the tourist attractions in town is the Buddhist *gompa* (see Fig. 4.7) high on a hill overlooking the valley. This *gompa* is being renovated, including the addition of a new guesthouse in the hopes that more trekkers and benefactors will come to visit now that the road is completed to Muktinath. When the senior author visited Jhong in April–May 2009 there was only one guesthouse, testament to how few trekkers came through the area in the past. The arrival of the Muktinath Road has brought the potential for many more guesthouses to be built in Jhong in the future.

4.4.5 Chungkhar

Chungkhar (see Fig. 4.7) is a small village located on the alternate trail between Muktinath and Jhong. The villagers interviewed in Chungkhar said they do not have many trekkers stay in their town because there are no guesthouses and their village is only an hour's walk from Muktinath, so most trekkers prefer to go on to Jhong. However, now that trekkers are starting to use the alternate trail through Chungkhar to Jhong one new guesthouse is under construction. As the number of trekkers increases there will be more land use changes of this nature in Chungkhar, and once the ring road is completed it may become a popular stop for new tourists arriving by road. In addition, there is a Buddhist monastery being built near Chungkhar and, as in Jharkot, the road is helping get materials to the construction site.

4.4.6   Putak

Putak (see Fig. 4.7) is a small enclave of houses one hour's walk down the valley from Jhong on the alternate trail. The beginning phases of a new ring road (discussed below) has brought land use changes to Putak. The road alignment is routed above the town, rather than following the traditional trail through Putak. Agricultural land was lost to the road, and it now bisects the remaining land. Currently, there are no guesthouses in Putak, however once the ring road is completed Putak has the potential to be another tourist destination in the valley. This would also potentially raise land values in the town and create land use changes by converting agricultural fields to guesthouses.

There are two new roads in the Muktinath Valley that developed in the 6 months between the two research periods in 2009 (April–May and October–November). The alternate trail down-valley from Jhong to Kagbeni had been upgraded to rural road status by widening and levelling it with bulldozers (see Fig. 4.7). Locals noted that this road would eventually connect to the Muktinath Road, essentially connecting the west side of the valley with the main road on the east side, making a ring road around the valley. Portions of this ring road were already passable by vehicles such as motorcycles and tractors.

The second new road is an extension of the ring road that leads from Chungkhar up the valley toward the Thorang La Pass, stopping at Thorang Phedi, the last set of guesthouses before the pass (see Fig. 4.7). A group of businesspersons built the road so that vehicles can supply the guesthouses.

Both of these roads have changed the land uses and the land values adjacent to the roads. The community plots in Jhong are substantially more valuable now that more trekkers are coming over from Muktinath to avoid the main road. In addition, once the ring road is completed (see Fig. 4.7), Jhong will be the first village on the opposite side of the valley to have a substantial number of guesthouses and therefore will benefit. The trail leading up to the Thorang La Pass has changed due to the new section of road built to Thorang Phedi. Once the road is good enough to transport tourists to the guesthouses, there will be a spurt of new building creating land use and land value changes.

4.4.7 Summary

Road construction has influenced the land use and land value of the villages in this case study. In Muktinath, there has been a threefold increase in religious pilgrims since the road was built, which has led to building a new pilgrims *dharamsala* in addition to a spur road to the temples. In Jharkot, the road caused a decrease in the number of trekkers, but it also provided a means for farmers to get their produce to market. This has led to a land use change through the expansion of fruit orchards. In Jhong, trekkers coming from Muktinath, to stay off the road, have influenced the villagers to divide a piece of community land into 64 one-acre plots for tourist development (see Fig. 4.7).

Two new roads were started in the 6 months between research periods in 2009. One, built on the alternate trail from Jhong to Kagbeni is part of a ring road that will circle the valley. Once completed, this ring road will bring many more tourists to the valley and land values will go up. Land use will change with portions of agricultural fields being replaced by tourist facilities. The section of trail from Kagbeni north to Lo Manthang will eventually be up-graded to rural road status. Once that section is completed there will be a road connecting Tibet with India through Nepal. Hence, land values along this stretch of road will increase dramatically with the large volume of trade that will be possible (see Fig. 4.7).

4.5   Case Study #2 the Marsyangdi Highway

In this case study we examine how road alignment can have a socioeconomic impacts on lodge owners in the villages of Bahundanda and Ghermu (Beazley 2013).

4.5.1 Bahundanda

The walk from either of the ACT starting villages of Besisahar (in the 1980s) or Khudi, to Bahundanda is a full first day on the trail for most trekkers (Fig. 4.8). The trail crosses to the east side of the Marsyandi River at Bhulbhule. In the past, after crossing the river, trekkers stopped at the old ACAP check post (Fig. 4.8) to show their trekking permits. Then, they climbed the long ascent to the town of Bahundanda, which can take several hours. However, upon reaching Bahundanda at the end of the day, the spectacular 360° panorama of the Himalayas made the climb worth the effort. For this reason, many trekkers stayed in Bahundanda before the new road arrived across the river at Sangye. After the road arrived, the old ACAP check post was moved from the east side of the Marsyangdi River to the Marsyangdi Highway side (west) of the river. Currently, all trekkers must stop at the new check post, and then many decide to walk up the new road or take a jeep to the end of the road at

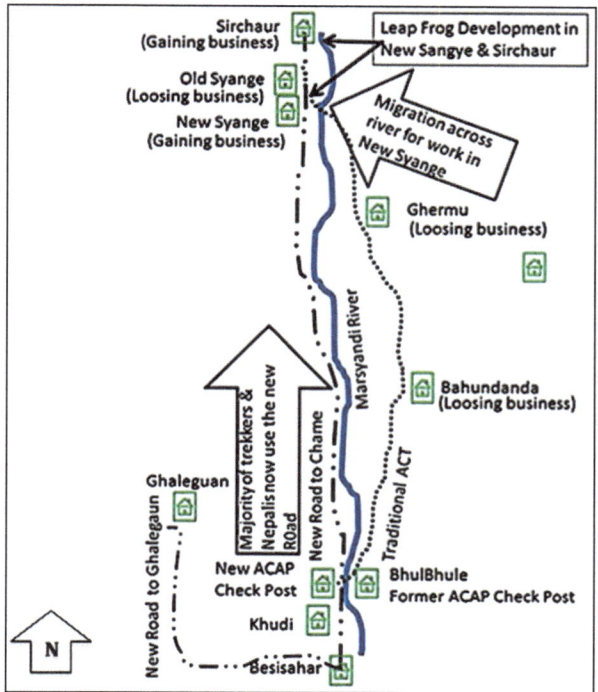

Fig. 4.8 Summary diagram of Case Study 2, the road alignment and the impacts on local villages (Beazley 2013)

Sangye rather than crossing the river and climbing the traditional ACT to Bahundanda. This makes for a much easier first day as the road is straight and climbs gently; hence trekkers can get further up the trail.

Many lodge owners in Bahundanda reported that their business is down due to the new road. Shop and restaurant owners whose clientele is both Nepalis and trekkers also confirmed this explaining that Nepalis choose to use the road for walking, for their mule trains carrying goods, or for a ride in a jeep, for the same reason as trekkers; it is flatter and easier than the tail.

This change in the flow of traffic is hurting the shop owners business in Bahnudanda, not only during the trekking season, but also in the off-season when they rely on income from Nepali customers. Several lodge owners maintained that the road caused them to lose 35–70% of their business. Most of the lodge owners agreed that they do not benefit at all from the road. They said that prices of goods and transportation have not gone down and it would be better if the road had never been built. Farmers interviewed all felt that the road was positive as it made their lives easier. Most of them claimed that price of goods and transportation had gone down.

Lodge owners in Bahundanda are using several different adaptive strategies to mitigate the impacts of the road on their businesses. One lodge owner who claimed his business was down 50% said: "The road is 99% bad and 1% good, 99% of the people on this side of the river don't want the road" (Shiva Pudrai, lodge owner, Bahundanda, personal communication, November 26, 2009). This contradicted several of the farmers and the health post officer, who all said they were very glad to have the road and added that if the road actually came right through Bahundanda, it would benefit them even more. Nonetheless, the lodge owner has put a sign in every room of his lodge explaining his views about the road. When farmers were asked about his statement that 99% of the people were against the road one farmer said: "Lodge owners are not everybody, local people like the new road. It saves us time. It used to take 2–3 days to go to Kathmandu, now we can get there in one" (Vijay Baad, farmer, Bahundanda, personal communication, November 26, 2009). This is a sentiment that was heard from farmers in virtually every village visited in 2009. Several lodge owners also made this point saying that villagers involved in tourism make up only 10–15% of the inhabitants and therefore the road is good for the majority.

Lodge owners in Bahundanda have tried several other adaptive strategies. It is interesting to note that these are both reactive and proactive strategies. They realize that there is no stopping the road at this point and consequently have tried several initiatives to influence trekkers to use the traditional trail through Bahundanda, including:

- Putting a sign at the new ACAP check post in Bhulebule pointing the way across the bridge to the traditional trail (ACT) that goes through Bahundanda.
- Putting a sign in guesthouse rooms about the road and why it is not fair.
- Requesting that the ACAP check post be relocated in Bahundanda so all trekkers will have to come to Bahundanda to have their permits checked.
- Requesting that jeeps going up the road only take Nepali customers.

According to one lodge owner, the signs have not made any difference so far, and the other requests were denied (Khrisna Shrestha, lodge owner, Bahundanda, personal communication, November 26, 2009). The one proactive strategy was to request the Nepal Tourism Board to develop another trekking trail from Bahundanda to a nearby scenic lake in honor of the famous Nepali conservationist, Dr. Harka Gurung.[2] Dr. Gurung grew up in Bahundanda and the lodge owners hope that the new trail would attract more trekkers because of his reputation.

These strategies highlight the fact that communities on the east side of the Marsyangdi River cannot ask ACAP for help. Technically, they are just outside the ACAP boundary formed by the Marsyangdi River (see Fig. 4.8). Many lodge owners in Bahundanda complained about ACAP. They felt that ACAP capitalized on tourism by collecting ACAP entry fees for many years when the old check point was outside the ACAP boundary (on the east side of the river; see Fig. 4.8). ACAP took the money and never did any community development projects in their village, or any other village on the east side of the river (Krishna Shrestha, lodge owner, Bahundanda, personal communication, November 26, 2009). The fee that ACAP collects from trekkers is supposed to go toward community development

---

[2] http://www.nepalitimes.com/issue/317/remembrance/12572.

projects within ACAP, but technically ACAP does not have to use any of that money for projects in communities on the east side of the river because they are outside the ACAP boundary.

Farmers' adaptive strategies for taking advantage of the road in addition to the obvious ones of getting produce to market faster, cheaper, and easier were as follows:

- One farmer's son is a jeep driver in Besisahar and transports goods and people along the new road.
- One of the *Dami* (occupational caste) villagers who previously walked with loads of fruit on her head from Besisahar to Bahundanda (to sell to trekkers and others in Bahundanda) can now bring bigger loads to Bahundanda with less walking by taking a jeep to the trail opposite Bahundanda, making her life much easier.

4.5.2   Ghermu

Ghermu is the next major village approximately 10 km up the trail from Bahundanda (see Fig. 4.8). It is smaller than Bahundanda and does not have Bahundanda's panoramic views, but it was popular with trekkers who wanted to go a little further than Bahundanda or, if the lodges in Bahundanda were full, it provided an alternative. There were no lodge owners in Ghermu to interview in 2009 because most of them had decided to move their bases of operation to the west side of the river in Sangye, due to the decrease in business in Ghermu after the road arrived (see Fig. 4.8). This is an adaptive strategy to take advantage of the opportunity the arrival of the road provided. All trekkers and Nepalis get off the bus at the end of the road in Sangye, which is the perfect location to start a trekking lodge, restaurant, or small store.

4.5.3   Summary

This case study shows how the physical alignment of road can have a socioeconomic impact on villages both on and off the road alignment. Bahundanda's lodge, restaurant, and shop owners all experienced a significant decline in their incomes from trekkers and Nepalis after the road was built due to the alignment of the road remaining on the west side of the Marsyangdi River. While some trekkers still choose to take the traditional trail through Bahundanda, the majority choose to stay on the west side of the river and use the road to get further up the valley. The relocation of the ACAP check point from the Bahundanda side of the river to be on the road alignment has further influenced trekkers to remain on the road and continue up the valley. Nepalis who travel up and down the valley also favor the road over the traditional trail because it is faster and easier. However for those villagers who are not involved in some type of business dependent on travellers along the trail the

road has made their lives easier by shortening the distance needed to travel on foot
to get supplies and making travel outside the area to places such as Pokhara and
Kathmandu faster and easier in spite of the fact that the road alignment is on the
opposite side of the river. The lodge owners in Ghermu have taken advantage of the
economic opportunity that the road has brought to the west side of the river by
relocating and opening shops near the bus stop area.

References

Allan, N. (1986). Accessibility and altitudinal zonation models of mountains. *Mountain Research and Development, 6*(3), 185–194.

Asian Development Bank (ADB). (2006). *Reaching the poor: Strengthening inclusive road development and management*. Retrieved from http://www.adb.org/PRF/prc-areawide-road-network.pdf.

Beazley, R.E. (2013). *Impacts of expanding rural road networks on communities in the annapurna conservation area, Nepal*. M.S. Thesis, Department of Natural Resources, Cornell University, Ithaca, NY.

Blaikie, P., Cameron, J., & Seddon, D. (2002). Understanding 20 years of change in west-Central Nepal: Continuity and change in lives and ideas. *World Development, 30*(7), 1255–1270.

Blaikie, P., Cameron, J., & Seddon, D. (1980). *Nepal in crisis: Growth and stagnation at the periphery*. Delhi; New York: Oxford University Press.

Blaikie, P., Cameron, J., & Seddon, D. (1977). *The Effects of Roads in West Central Nepal, Part I (Summary)*. A report to the Economic and Social Committee for Overseas Research, Ministry of Overseas Development. East. Agnila, UK: Overseas Research Group, University of East Anglia.

Blaikie, P., Cameron, J., Feldman, D., Fournier, A., & Seddon, D. (1976). *The effects of roads in west Central Nepal. A report to the economic and social Committee for Overseas Research, Ministry of Overseas Development*. East Angila, UK: Overseas Development Group, University of East Anglia.

Bohle, H., & Adhikari, J. (1998). Rural livelihoods at risk how Nepalese farmers cope with food insecurity. *Mountain Research and Development, 18*(4), 321–332. doi:10.2307/3674097.

Brown, S. (2003). Spatial analysis of socioeconomic issues: Gender and GIS in Nepal. *Mountain Research and Development, 23*(4), 338–344.

Calvo, C. M. (1998). *Options for managing and financing rural transport infrastructure*. World Bank Technical Paper No. 411. Washington, DC: World Bank. Retrieved from http://www.ingentaconnect.com/content/wb/295/1998/00000001/00000001/art00001.

Cook, C., Duncan, S., Anil, S., & Wu, G. (2005). *Assessing the impact of transport and energy infrastructure on poverty reduction*. Manila: Asian Development Bank. Retrieved from http://www.adb.org/Documents/Reports/Assessing-Transport-Energy/assessing-transport-energy.pdf.

Hettige, H. (2006). *When do rural roads benefit the poor and how? Operations evaluation department*. Manila, Philippines: Asian Development Bank. Retrieved from http://www.adb.org/Documents/Books/ruralroad_benefits/rural-roads.pdf.

Howe, J. (1984). Chapter 3: The impact of rural roads on poverty alleviation: A review of the literature. In J. Howe & P. Richards (Eds.), *Rural roads and poverty alleviation* (pp. 48–81). Boulder, Colorado: Westview Press.

Ives, J. (2004). *Himalayan perceptions: Environmental change and the well-being of mountain peoples*. New York: Routledge.

Jacoby, H. (2000). Access to markets and the benefits of rural roads. *The Economic Journal, 110*, 713–737.

Kafle, M. (2007). *Nepal: Enhancing poverty reduction impact of road projects*. Asian Development Bank (ADB) Report. Retrieved from http://www.adb.org/Documents/Produced-Under-TA/39525/39525-NEP-DPTA.pdf.

Kreutzman, H. (1991). The karakoram highway: The impact of road construction on mountain societies. *Modern Asian Studies, 25*(4), 711–736.

Leinbach, T. (1995). Transport and third world development: Review, issues, and prescription. *Transportation Research Part A: Policy and Practice, 29A*(5), 337–344.

Paudyal, D. P. (1998). *Access improvement and sustainable development–rural road development in Nepal*. Kathmandu: ICIMOD. Retrieved from http://books.icimod.org/demo/uploads/ftp/Access%20Imporovement%20and%20Sustainable%20Development.pdf.

Phalebus, P. (2015, December 19) *Blacktopping of Kagbeni-Muktinath road begins*. The Kathmandu Post. Retrieved from http://kathmandupost.ekantipur.com/news/2015-12-19/blacktopping-of-kagbeni-muktinath-road-begins.html.

Richards, P. (1984). The economic context of rural roads. In J. Howe & P. Richards (Eds.), *Rural roads and poverty alleviation*. Boulder, Colorado: Westview Press.

Salick, J., Yongping, Y., & Amend, A. (2005). Tibetan land use and change near Khawa Karpo, eastern Himalayas. *Economic Botany, 59*(4), 312–325.

Schroeder, M. (1971). *Impact of the Sonauli-Pokhara Highway on the regional income and agricultural production of Pokhara Valley, Nepal*. (Thesis H36 1971 S381) Ithaca, NY: Cornell University.

Seddon, D., & Shrestha, A. (2002). Gender in rural transport development: Chattra Deruali, Nepal. In P. Fernando & G. Porter (Eds.), *Balancing the load: Women, gender, and transport* (pp. 235–245). New York: Zed Books.

Strickland, R. (2009). *External review of district roads support programme (DRSP)*. Retrieved from http://www.drspnepal.org/drsp/downloads/External-Review-DRSP-2009.pdf.

Van de Walle, D. (2008). *Impact evaluation of rural roads project*. World Bank Report. Retrieved from http://siteresources.worldbank.org/INTISPMA/Resources/3837041146752240884/Doing_ie_series_12.pdf.

Chapter 5
Sociocultural Impacts of Roads

Abstract There is a paucity of empirical research on the sociocultural impacts of roads. Even though there is a growing acknowledgement by donors, INGOs, NGOs, and government institutions that this is an important component of infrastructure planning, development, and implementation, sociocultural impacts are hard to quantify. Some sociocultural impacts may not manifest until years after the project is complete. This situation is greatly exacerbated in culturally diverse countries such as Nepal. In Nepal this has lead to instances of theft, illegal trade and smuggling, violence, and community conflicts. This was documented on both the Marsyangdi Highway and Kali Ghandaki Highway where changes to cultural traditions, intentional road blockades, and social unrest were all observed.

Keywords Sociocultural impacts • Cultural tradition • Hinduism • Tibetan Buddhism • Pilgrimage • Sacred sites • Strike (*bhanda*) • Maoist • Annapurna conservation area • Kali Ghandaki highway • Marsyangdi highway • Hindu caste groups • Ethnic group • Trade • Tourism • Trekking • Mobility • Migration • Manakamana • Muktinath

5.1 Overview of Sociocultural Impacts of Roads

There is much less literature on the sociocultural impacts of roads. While the need for this type of assessment is increasingly recognized as important by governments, donors, and non-governmental agencies, sociocultural impacts are hard to quantify, and certain ones may not show up until long after the road project has been completed. This situation is greatly exacerbated in culturally diverse countries such as Nepal, which has 103 distinct caste and ethnic groups with 92 different living languages (Gurung et al. 2006). In this chapter, we explore the implications of building roads into culturally diverse and previously roadless areas and the resultant changes in sociocultural traditions that may occur through analyses of several case studies as well as our empirical research on the Kali Ghandaki and Marsyangdi Highways.

© The Author(s) 2017
R.E. Beazley, J.P. Lassoie, *Himalayan Mobilities*, SpringerBriefs
in Environmental Science, DOI 10.1007/978-3-319-55757-1_5

One set of sociocultural impacts that has been well documented is the increase in the sex trade, HIV, and sexually transmitted infections (STI) associated with roads increasing mobility and the number of migrants (ADB 2008; Brushett and Osika 2005; UNDP 2006). These diseases are often called 'highway diseases' due to the well-established connection between the commercial sex trade and roads. A report by the International Labor Organization cites two studies that document the linkage between roads and HIV. One study, conducted along a major highway in southern India, showed that 16% of the drivers were found to be HIV positive versus the national average, which was less than 1%. The other study, in the KwaZulu/Natal Midlands of Africa conducted in 2001 by the South African Medical Research Council, found that 56% of long-distance truck drivers in the region were HIV-positive (ILO 2005). A World Bank report in 2008 estimated that along the major east-west transport corridor in West Africa (Abidjan-Lagos transport corridor), which connects five countries and where three million people cross borders each year, there could be as many as 300,000 HIV infected people travelling along the corridor annually (Brushett and Osika 2005). Sex is not the only source of HIV as drug use is often also associated with the spread of HIV along road corridors (ADB 2008). Mobility increases people's ability to interact and engage in activities that can lead to high risk pursuits contributing to increased health problems and socio-cultural changes in attitudes, values, and lifestyles.

A study in the Brazilian Amazon by Ayers et al. (1991) showed significant sociocultural changes as the result of new road construction. The village they surveyed was isolated from any road until 1978. The first survey was done just before arrival of the road in May 1978 and the second was done in May 1980. While the primary purpose of the study was to assess changes in subsistence hunting, the interviews and data collected showed some interesting overall sociocultural changes that were attributed to the road. One of the biggest changes was that the population size of the town increased by almost 55% as a result of immigration from the southern and eastern parts of the country. Land use and agriculture practices changed as locals discovered that land value correlated with crops, and to maintain possession of their land they needed to keep it under constant cultivation. Linkage with the national economy brought new food sources that changed peoples' diets and their traditional livelihood strategies. Before the road, the main source of protein in diets was from animals that were hunted in the area. With the arrival of the road came access to fresh beef, which was a more reliable source of protein than hunting, resulting in an increase in beef consumption and less reliance on hunting. Other economic activities that came with the road that affected people's traditional livelihoods included large-scale agriculture, cattle ranching, mining, and logging. (Ayers et al. 1991).

Perhaps the most well-known road project that created numerous negative impacts, not only in the sociocultural sphere but socioeconomic and environmental spheres as well, is the paving of the BR-364 Highway in Brazil. This project began in the 1980s as part of the Northwest Regional Development Program (a.k.a. *Polonoroeste*) in Brazilian Amazonia. The project, partially funded by the World Bank, was supposed to "… provide an integrated approach to frontier development and avoid further land conflicts and illegal logging" (World Bank 2006: 50). The reconstruction and paving of BR-364 and additional feeder roads, while only part of the project, is largely thought to be responsible for the ensuing rapid deforestation and social conflict (World Bank 2006; Peet 2003; Keck 1998; Price 1989; Fearnside 1987; Lutzenberger 1985). The paving of the highway led to a mass influx of migrants looking for new opportunities with the population of the area almost tripling in 6 years, growing from 600,000 to 1,600,000 (Peet 2003). Sociocultural impacts included armed conflict between squatters and landowners including indigenous inhabitants over land rights (Brunelli 1986), displacement of indigenous inhabitants (Price 1989), frequent abandonment of homesteads (Keck 1998), unscrupulous land profiteering (Price 1989), encroachment on Amerindian reserves (Feranside 1987), and thousands of deaths from malaria (Peet 2003).

Several studies have also been done in China concerning the impacts of roads on ethnic minorities. For example, Salick et al. (2005) found that traditional agricultural practices in Tibetan villages near new roads in Northwest Yunnan had changed. Villages near roads grew significantly more cash crops, such as grapes and wheat, compared to villages not near roads, which tended to grow more barley and buckwheat. In addition, the traditional soil fertilizers of oak leaves and dung were being supplemented with chemical fertilizers, and pesticides were also being used. Traditional livelihood practices had also changed due to roads with males leaving villages to find cash income on road construction crews and elsewhere. In one village, road construction had reduced its water supply resulting in the abandonment of rice cultivation in the area.

In addition, other studies have looked at the impact transportation, especially roads and railroads, has had on the migration of Han Chinese and the Chinese military to ethnic minority areas. The resulting socioeconomic and sociocultural changes have often led to violent conflict and marginalization of local inhabitants in areas such as the Tibet (Dreyer 2003) and Xinjiang Uyghur Autonomous Regions (Becquelin 2000).

Cruikshank (1985) analysed the impacts of the Alaska Highway (AH) on the First Nation inhabitants of Canada. She found a substantial migration in the First Nation population as the focus of transportation changed from river to road during the construction of the AH (March 8–September 24, 1942). First Nation people began to abandon river communities where they had formerly traded furs based on riverboat transport. Once riverboats stopped running, they moved closer to the road, not only to take advantage of the new system of transportation for trading goods, but also to seek additional employment either working on the road or in other commercial activities. These new areas near the road could not support their traditional subsistence lifestyles, so they had to seek wage labor to survive.

The First Nation traditional hunting patterns were also affected by creation of the Kluane Game Sanctuary (December 1942) along a section of the AH, which eliminated their hunting rights. In another area, competition from non-native road workers who had been granted hunting privileges by the government, affected the First Nation hunters. Many of non-natives hunted for sport, leaving the dead animal to rot without taking the meat, which the First Nation saw as wasteful and counter to their ethical code of respect for the land and animals that provide food.

As the result of a large influx of male workers and army personal during the construction of the AH, negative impacts on the Canadian First Nation people such as changes in traditional values and social behavior, increase in alcohol use, and health epidemics such as measles, whooping cough, and meningitis ensued.

"Before the highway came and split us all in different ways, we used to feed ourselves good from this country" (Yukon First Nation person, quoted in Cruikshank 1985: 178). This quote reflects the breakdown of many of the traditional First Nation ways, which they identify with the coming of the highway. Benefits did accrue after the completion of the highway, such as access to medical facilities and better communication services. However, in terms of the First Nation population, Cruikshank (1985: 185) points out that: "The highway was a decisive factor bringing Yukon Indians to the marginal position they have in the present Yukon economy and society."

Other sociocultural impacts of roads have been reported in terms of gendered mobility. They are not as easily quantifiable because culturally gendered norms of behavior in different populations influence how both men and women access and use forms of mobility. Case studies show that interpretation of religious edicts as well as culturally embedded gender roles have a significant influence on gendered mobility. Better access to mobility brought by new and improved roads in some cases eases women's work load but in other cases increases it as men use their new mobility to seek wage labor away from home (see Chap. 2, Sect. 2.2 Gendered Mobility).

5.2   Sociocultural Impacts of Roads in Nepal

A long-term study of the Hindu sacred site at Manakamana, which sits on the top of a hill in central Nepal, credits the road in the valley below, which first arrived in 1976, for the beginning of the "… large-scale commercialization of goods and services for pilgrims" (Bleie 2003: 181). A cable car was built from the road to the temple in the late 1990s, eliminating the need to make the 'arduous' journey up the hill on foot and increasing the number of pilgrims even more. While the local community enjoyed a temporary economic benefit from the 250,000 visitors who came to Manakamana annually, the large increase in numbers created environmental problems that the villagers had not anticipated and were unprepared to deal with, such as the increased demand for wood for cooking, limited drinking water, and waste removal. Eventually, urban developers, including the cable car company, built larger, modern food and lodging facilities. This took a large piece of the locals' share in the economic profits leaving the village with the entire burden and few benefits.

The local community experienced other unexpected consequences of the increase in pilgrims. In 1999, heavy monsoon rains resulted in water flooding into the temple courtyard damaging the foundation and almost inundating the inner shrine. Some villagers attributed this to the huge number of pilgrims whose non-traditional manner offended the temple Goddess, and they feared she might leave. Others blamed the construction of the cable car, as it had ruined the pre-existing drainage system. Either way, the cable car was to blame as no one in the village could recall a time in the past when this calamity had occurred (Bleie 2003).

Mobility is not just a physical phenomenon. In Nepal, which was a Hindu monarchy from the mid eighteenth Century until 2008, the Hindu caste system influenced mobility physically, economically, and socially. Increased mobility has spurred the trend to seek wage labor in the Gulf countries that Nepal has been experiencing for the last several decades (ADB 2010; Liechty 2003). This physical mobility often leads to greater socioeconomic mobility, which can also increase sociocultural mobility even in the case of the most marginalized, such as Dalits (formerly called untouchables). In some communities in Nepal every Dalit household has sent a worker abroad. Remittances from this work abroad have brought positive changes in health, hygiene, physical infrastructure, and education (Gautam 2014).

Case studies on gendered mobility in Nepal reveal additional sociocultural impacts such as the increase in female trafficking and the commercial sex trade, as well as the spread of sexually transmitted diseases. Increased mobility has the potential to increase income for rural agricultural communities through better connection to markets. However, market demands often require change from traditional cropping and animal husbandry practices to more intensified and less diverse practices, which has been shown to increase women's workloads. Positive sociocultural benefits such as better connection to relatives and friends, easier access to both religious sites and teachers, and increased educational opportunities have also been reported (see Chap. 2 Mobilities, Sect. 2.2.1 Gendered Mobility in Nepal).

5.3 Summary

One of the main livelihood strategies that roads facilitate is in and out-migration. This can have both positive and negative impacts on sociocultural elements. Negative impacts, such as changes to traditional sociocultural values and practices, often result from the influx of migrants to areas newly opened by roads. HIV/AIDS and sexually transmitted infections, sex trafficking, and drug use has been shown to have a link to road connectivity. Out-migration of family members (usually male) to seek employment, while often having a positive socioeconomic impact, can also lead to increasing gender inequality by placing more of the burden of household chores on women. Conversely, in some areas it was noticed that increased mobility could lead to men shouldering some of the tasks previously done by women, thereby relieving some of the gender inequality (see Chap. 2, Sect. 2.2). When new roads are constructed into extremely isolated areas, they facilitate a large population influx of outsiders, which can lead to the marginalization of native inhabitants.

Positive sociocultural outcomes have been observed in terms of ease of access to other areas and the resulting social capital, in addition to increased access to employment, schools, and hospitals. In some cases, empowerment of women and marginalized groups also has been noted.

5.4 Case Study #1 The Kali Ghandaki Highway

This case study investigates how the Kali Ghandaki Highway is influencing the traditional Himalayan pilgrimage (Beazley 2013).

5.4.1 Background

After the construction of the new Kali Ghandaki Highway and the extension to Muktinath in 2008 (Fig. 5.1), the number of pilgrims arriving at the sacred site at Muktinath increased considerably. The head Hindu priest at the Muktinath temple said the numbers had increased "radically" by two to three times compared to before

Fig. 5.1 Location of the villages of Muktinath, Jharkot, Jhong, Chungkhar, and Putak in the Jhong Khola River Valley in Mustang District (Adapted from Google Earth)

the road (Shiva Ram Shuba, Head Hindu priest, Muktinath, personal communication, November 27, 2009). The Buddhist nuns, who are the traditional caretakers of the temples at Muktinath, also said both Hindu and Buddhist pilgrims increased after the road reached Muktinath (Tezing Tsyeyang, Buddhist nun, Muktinath, personal communication, November 28, 2009).

5.4.2   Muktinath

Several changes have occurred because of this increase in the number of pilgrims. *Muktichettra* is the Sanskrit name used for Muktinath. In Sanskrit, *mukti* means salvation and *chettra* means field. *Chettra* can mean field in two ways. One is secular as in a field around commerce such as tourism and trade, and the other *chettra* is sacred, as in the spiritual field that encompasses sacred objects (Messerschmidt 1992). With the arrival of the road at Muktinath, there is now a steadily increasing overlap of these two *chettras,* with commercialization of the town and the temple area increasingly evident. This is due to the new road, which makes access to the sacred site easier for everyone. The most obvious change at Muktinath is the constant line of motorcycle taxis (Fig. 5.2) that take pilgrims from where the jeeps stop, through the village and up the hill to the temples; a distance of about 1 km. This situation is partially due to the fact that many pilgrims arrive at the high elevation of

Fig. 5.2 Motorcycle taxis taking pilgrims to the temples at Muktinath (R. E. Beazley 2009)

Muktinath (3800 m), from lower elevation locations, such as Pokhara (820 m), in one day without time to acclimatize, now that the road is complete. Before the road they had time to acclimatize by walking to this elevate over a 5–7 day period.

We argue that this has resulted in a fundamental cultural shift in values of pilgrims. In the traditional Himalayan pilgrimage, the journey is equally, if not more, important as the destination. The hardship of the journey is the means by which the pilgrim can experience a spiritual awakening that then transforms the pilgrim into a receptive vessel (Huber 1999). Once the pilgrim reaches the sacred site, due to this spiritual awakening in route, they are then empowered to engage with the sacred field and the blessing bestowed by its power, the power of the deities who inhabit it, and by their symbolic representations (Sax 1990). We suggest that traveling by vehicle removes many of the challenges and hardships of the journey and therefore the destination becomes the goal. In this case, the pilgrim relegates the spiritual transformation to the external field and its objects rather than the catalytic internal transformation in conjunction with the external field. According to Bharati (1981: 5–6 quoted in Sax 1990: 492), the renowned Indian anthropologist and expert on Hindu pilgrimage:

> In all Indian languages, Sanskritic or Dravidian, the word for pilgrimage contains the root for 'to go,' 'to move' (*lgam-, gaccha-, yam, yaccha-, yatra*)... [I]t was the motion, the effort of moving on and then up the steep mountain, the circumambulation and the various prostrations in and around the shrine ... which must be seen as the key element in the whole undertaking.

This is not to say that all pilgrims enduring hardship experience a spiritual transformation or that pilgrims who come by vehicle do not experience a transformation, but it is a cultural shift in the traditional values and mode of pilgrimage in the Himalayas. Bleie (2003: 177) states this succinctly, saying: "... the increasing use of mechanical transport to pilgrim sites in the Central Himalayas erodes the cultural notions that have underpinned the Himalayan pilgrimage for centuries." She (Bleie 2003: 179) goes on to say:

> Moreover, incipient transformation in the basic cultural categories of worship, place, and journeying of Nepalese and Indian pilgrims also blurs the boundary between pilgrimage and tourism. It is principally the pilgrims' use of modern means of transportation that spearheads this cultural change.

In terms of commercialization, it could be argued that commercialization of sacred sites is not a new phenomenon. In the Himalayas, the close historical link between pilgrimage, trade, and festivals near sacred sites undoubtedly led to a type of commercialization. The difference the road brings is a speeding up of the process and an increasing sophistication of the change. One of the significant changes is that previously only devout practitioners would undertake a strenuous journey by foot to a sacred site. With the ease and speed of vehicular travel, not just the devout take the journey. Various other individuals with ulterior motives arrive, such as thieves who want to steal sacred objects to sell on the black market or tourists who do not respect or understand the sacred space and are unaware that they create a market for defamers, charlatans, and voyeurs, which creates an atmosphere of corruption. The difference in wealth between the tourists and the local inhabitants, including holy men and women, leads to temptations that ultimately erode traditional morals and ethics.

There are currently plans to build a cable car to the sacred site at Muktinath (GoN/MoF 2009) and a feasibility study has been done ("Two New Cable Cars" 2009). One can only guess what the impact will be. However, if it is anything like the changes at Manakamana, we believe that Muktinath will undergo a vast and profound transformation from an ancient sacred site to a commercial tourist attraction. In addition to the cable car there is a road planned to link the main road in the Kali Ghandaki River Valley, which currently leads to Muktinath, with the road that connects the Tibetan border to Lo Manthang in Upper Mustang (NTNC 2008) (see Fig. 5.3). With the completion of this section of road there will effectively be a road connecting Tibet with India through Nepal. This link existed a long time ago with the Trans Himalayan Salt Trade and, depending on the geopolitical climate, full-scale trade through the Kali Ghandaki River Valley may begin again. If it does, the number of vehicles will increase substantially bringing more pollution and more people. There is already concern that the road will bring looters seeking artifacts from archaeological and sacred sites to sell on the black market (Turner 2010).

The positive side is the potential for the road to enable the local people to raise their standards of living with better access to a number of social and economic benefits, including access to markets and other job opportunities, schools, hospitals, and extended family members.

Fig. 5.3 Kali Ghandaki Highway (*red*) with Muktinath spur road (*yellow*) (adapted from Google Earth)

5.4.3 The Muktinath Valley Ring Road

A similar sociocultural change may develop in the future after the Muktinath Valley Ring Road (MVRR) around the Muktinath Valley is completed (Fig. 5.4).

There is a traditional Tibetan Buddhist *kora* or circumambulation, of the six villages of the Muktinath Valley. This *kora* is called *Chokhor Kora* or the "the scripture circuit" because all 108 volumes of the Buddhist cannon called the *Kangyur* are carried around the valley to bless the space (Gutschow 1999). This tradition is also practiced on an individual village basis to bless the fields and ensure a good harvest. Once the MVRR is complete, we suggest that this tradition will change from foot travel to vehicular travel.

This use of vehicles for pilgrimage is a growing trend in Tibetan areas, as more roads are constructed into areas where previously there were only trails (Buffetrille 2003). The senior author also observed this in July and August of 2009 in China along the Kawagebo Kora, on the border of Northwest Yunnan and Tibet.

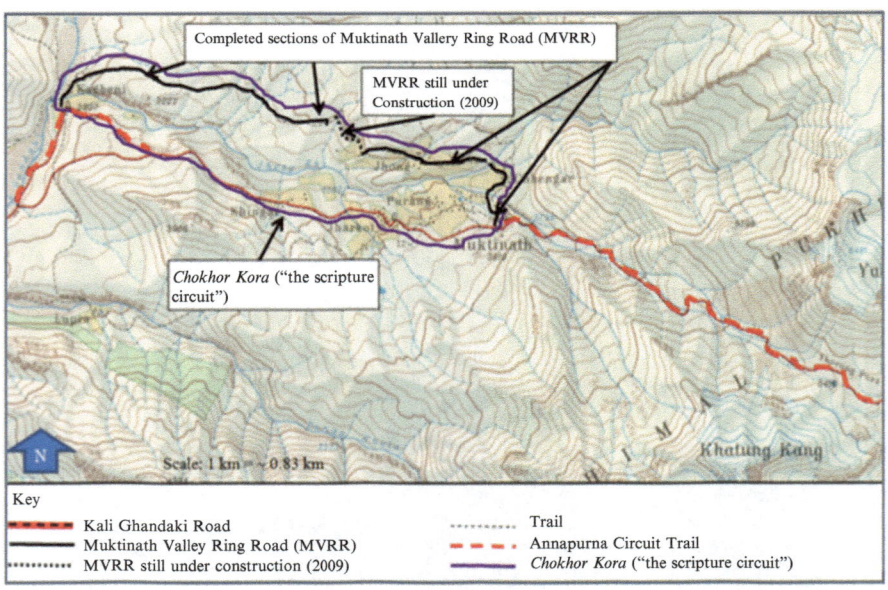

Fig. 5.4 Map of Muktinath Valley with the Muktinath Valley Ring Road, the *Chokor Kora*, and trails (Adapted from Nepal-Kartenwerk der Arbeitsgemeinschaft für vergleichende Hochgebrifsforschug Nr. 9, 1993) (Beazley 2013)

5.5   Case Study #3 Unintended Consequences of Road Development

This case study considers two villages, Tal and Tukche, which share something in common in relation to the road even though they are spatially distant from each other, with the whole Annapurna Massif between them and are on different sides of the Thorang La Pass (Fig. 5.5) (Beazley 2013). Specifically, it highlights how roads can have an impact that is usually unanticipated by planners, policy makers, or government officials, and illustrates another important point about roads—they can be used to both facilitate movement and to stop it.

5.5.1   Background

Roads are often only thought of terms of economic development. However, roads are also an instrument of gaining and establishing control of an area. There have been many roads built specifically for security forces or military campaigns. The Roman Roads (Laurence 1999), the Imperial Chinese Roads (Needham 1971), the Burma Road constructed during WW II (Tuchman 2001), and the Karakoram Highway (Kreutzman 2004), and are but a few examples.

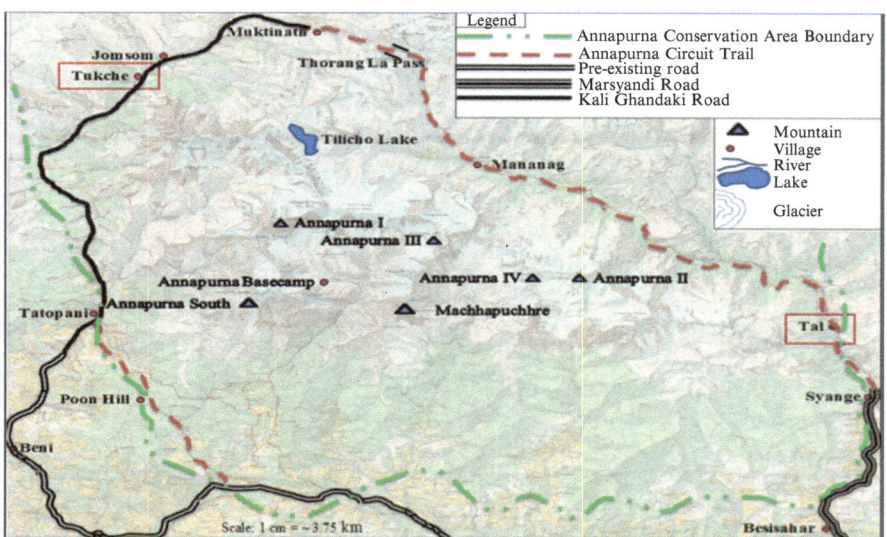

Fig. 5.5 Villages of Tal and Tukche in the Annapurna Conservation Area (Adapted from Nepal-Kartenwerk der Arbeitsgemeinschaft für vergleichende Hochgebrifsforschug Nr. 9, 1993) (Beazley 2013)

Kreutzman (2004: 204–205) points out that the strategy of blocking roads for political reasons has been used extensively in many mountainous areas: "Blockages of the Karakoram Highway have been used by inhabitants of Kohistan in May–June 1993 as a political tool to convince the public administration that timber exports from the few remaining, and rather depleted, natural forests should no longer be prohibited." Blocking roads in mountainous areas is especially effective. Due to the terrain, there is often no alternate route to bypass the blockade. For centuries, military campaigns have effectively used roadblocks to lay siege to cities they wished to capture. In fact this is exactly how the King of Gorkha, Prithvi Narayan Shah captured the Kathmandu Valley- blocking all the trails leading into the city effectively laying siege and thus became the first king of unified Nepal (Whelpton 2005). Roads are also areas of vulnerability to attack from marauding bandits, thieves, terrorists, and counter insurgents.

In Nepal, strikes (*bhandas*) are the most popular form of protest. This usually involves blocking a road to traffic in some manner (Fig. 5.6). *Bhandas* are extremely effective in Nepal because there are only a few roads that service the capital, Kathmandu. If one of the roads is blocked it can have a major influence on commerce in the city. When this happens the government is forced to take action to appease the protesters with some type of compromise. Consequently, *bhandas* have become an effective form of civil protest and negotiation for the public since the first pro-democracy People's Movement demonstrations in 1990 (Shah 2007). During the People's War (1995–2006), control of different sections of roads in Nepal was crucial in gaining control of an area for both the Royal Nepalese Army (RNA) and the Maoist People's Liberation Army (MPLA).

Fig. 5.6 A *bhanda* on the road between Kathmandu and Pokhara. The villagers overturned a truck carrying potatoes to block the road in protest (R. E. Beazley 2009)

As the Maoist moved from their headquarters in western Nepal east toward Kathmandu, they destroyed bridges and other infrastructure to disrupt the government and limit the movement of the RNA and police. After a bridge, section of road, or building was destroyed, the MPLA would flee the roads and move into the mountains on trails. Unfortunately, this did not just affect the RNA but it also disrupted the lives of local people, making it dangerous and difficult to travel. The MPLA hid troops in the Annapurna Conservation Area (ACA) where they carried out guerrilla warfare tactics against the Annapurna Conservation Area Project (ACAP) and other institutions associated with the monarchy.

Security became an issue for anyone working and living in the remote mountainous areas, particularly in ACAP (Baral et al. 2008). A bomb was detonated at ACAP headquarter in Pokhara and two ACAP officials were lynched in Ghandruk and the ACAP headquarters there was destroyed (Murphy et al. 2005). During this period many Nepalis left Ghandruk for safer areas and all the guesthouses were mostly empty. This was a feather in the cap of the Maoist because Ghandruk was the first village established in ACAP as pilot project community and is ACAP's pride and joy. ACAP's connections to the monarchy ensured that this attack was viewed politically as a decisive victory for the Maoists. This is only one of the numerous incidents that happened during the People's War, not only in ACAP but also in many mountain communities across Nepal.

As a result of this escalating turmoil, King Gyaendra declared a state of emergency in 2006 giving him complete control of the RNA. He ordered the RNA to go up the Kali Ghandaki Valley and complete a section of the road near Marpha to take control of the area.

These incidents highlight one of the most important aspects of trails and by extension roads. Whoever controls a given section of a road or trail influences the mobility of the people who are in the vicinity of that section of road and those that rely on the road for transport. This aspect of control is illuminated below by comparing two villages were locals blocked the trail/road to protest what they considered unfair treatment. We posit that roads are inherently political in nature and are embedded with issues of power and control.

5.5.2   Tukche (Mustang District)

Just before the senior author arrived in the Kali Ghandaki Valley in the spring of 2009 there had been a *bandha* (protest/roadblock) at Tukche (see Fig. 5.5). In this case, the villagers to protest unfair bus and jeep ticket sales further up the valley in Jomsom put the *bandha* in place. The only airport in the area is located in Jomsom. Consequently, many tourists fly in and out of Jomsom. After the Kali Ghandaki Road was completed, trekkers began using trucks, jeeps, and bus transportation to get out of the valley rather than flying or hiking out. Transport operators charge foreigners more for a seat than they charge Nepalis, so they prefer to take foreigners. As a result, many of the vehicles were full by the time they reached Tukche, about 20 km down valley from Jomsom. Ironically, even though the people of

Tukche finally had road access to their village, it did them no good because the vehicles traveling the road were most often full and would not stop. The villagers decided to take matters into their own hands and tried negotiating with the transportation operators, but with no success. They realized that the only way to affect a change was to block the road to traffic. The *bandha* forced the transportation companies to comprise and agree not to sell all the seats in Jomsom and to arrive in Tukche at scheduled times to pick up villagers (Laxmi Bista, lodge owner, Tukche, personal communication, November 30, 2009).

The senior author also interviewed several guesthouse owners in this area who said their business had been down during The People's War. This was due to Maoist-led incidents, such as trekkers being asked for donation at gunpoint, kidnappings, and the closing of one sections of the Kali Ghandaki Road (see Fig. 5.1), preventing trekkers and locals from traveling up and the down the valley (Sarita Gurung, lodge owner, Tukche, personal communication, November 30, 2009).

5.5.3 Tal (Manang District)

The case of Tal (see Fig. 5.5) is similar to Tukche in that the villagers blocked the ACT to protest what they considered unfair treatment, in this case in terms of the road alignment. The road was not yet completed to Tal because the section of the Marsyangdi Valley leading up to Tal is vertical with numerous cliffs making road construction very difficult. The ACT is on the same side of the river as Tal, therefore anyone traveling on the trail goes through village. The current road alignment however, is planned to stay on the opposite of the river from Tal. The villagers are concerned that they will lose all their business because vehicles will completely bypass their village on the opposite side of the river. To address this issue they blocked the ACT on either end of their village, allowing only trekkers to pass. After a week, the government sent the Minister of Transport by helicopter to Tal to negotiate. The result was a promise by the government to build a spur road that would cross the river, pass through Tal, and then cross back to the main road (Ram Bahadur Regmi, lodge owner, Tukche, personal communication, October 28, 2009).

There are also plans to build a dam and a hydroelectric plant not far from Tal. This, of course would necessitate a road because using helicopters to fly equipment and material to the site is too expensive. In anticipation of this project, several villagers are constructing new guesthouses in hopes of renting them on a semi-permanent basis to the construction company (Ram Bahadur Regmi, lodge owner, Tukche, personal communication, October 28, 2009). If the hydroelectric project does actually begin it is possible that Tal will have more than one road.

5.5.4 Summary

We argue that an additional impact of road construction is the potential for them to become pawns in political struggles and civil rights issues. The state uses roads to extend its control and protect national boundaries. However, this is only true to the extent that the state has control of the roads. In Nepal, civil society and the MPLA have found that once they have control of a section of road they have power to leverage the state.

It is ironic that roads are originally designed to facilitate the flow of people and goods, but they can also be used to stop the flow of traffic in order to force a negotiation, such as in Tal and Tukche. We posit that power is a central issue with roads as demonstrated in this case study. Whoever controls a section of road to a large part also controls the people near the road. Gaining control of a road can be accomplished by blocking it to traffic, building a new road that connects to pre-existing roads, or by destroying bridges and sections of road. When a road is constructed virtually all the adjacent vegetation is leveled, and in some cases just beyond this zone there are either trees or hills where groups or individuals wishing to attack travellers can hide. This land use change makes roads one of the most vulnerable spaces to attack and gain control.

In both Tukche and Tal (see Fig. 5.5) road construction has led to a sociocultural change in the form of protest in which the villagers used the road as a political tool to fight for equality. It is instructive in the above case to notice that a socioeconomic motivation, the ability to use the road for commerce, led to a sociocultural change, protest, which may lead to another environmental change in the case of Tal if the spur road is ever built.

References

Asian Development Bank (ADB). (2010). *Key indicators for Asia and the Pacific 2010, special chapter the rise of Asia's middle class*. Asian Development Bank. Retrieved from http://www.adb.org/publications/key-indicators-asia-and-pacific-2010.

Asian Development Bank (ADB). (2008). *ADB, Roads, and HIV/AIDS: A resource book for the transport sector*. Retrieved from http://www.adb.org/Documents/Books/ADB-HIV-Toolkit/ADB-HIV-Toolkit.pdf.

Ayers, J., De Magalhes Lima, D., De Souza Martins, E., & Barrieros, J. (1991). On the track of the road: Changes in subsistence hunting in a Brazilian amazonian village. In J. G. Robinson & K. H. Redford (Eds.), *Neotropical wildlife use and conservation* (pp. 82–92). Chicago: University of Chicago Press.

Baral, N., Stern, M., & Bhattarai, R. (2008). Contingent valuation of ecotourism in annapurna conservation area, Nepal: Implications for sustainable park finance and local development. *Ecological Economics, 66*(2–3), 218–227.

Beazley, R. E. (2013). *Impacts of expanding rural road networks on communities in the annapurna conservation area, Nepal*. M.S. Thesis, Department of Natural Resources, Cornell University, Ithaca, NY.

Becquelin, N. (2000). Xinjiang in the nineties. *The China Journal, 44*, 65–90.

Bleie, T. (2003). Pilgrim tourism in the central Himalayas the case of Manakamana temple in Gorkha, Nepal. *Mountain Research and Development, 23*(2), 177–184.

Brunelli, G. (1986, June 30). Warfare in polonoroeste. *Cultural Survival Quarterly,* 10(1981–1989), 37.

Brushett, S., & Osika, J. S. (2005). *Lessons learned to date from HIV/AIDS transport corridor projects*. World Bank Global HIV/AIDS Program. Retrieved from http://siteresources.world-bank.org/INTHIVAIDS/Resources/375798-1103037153392/Transport.pdf.

Buffetrille, K. (2003). *The evolution of a Tibetan pilgrimage: The pilgrimage to a myes rMa chen mountain in the 21st century*. 21st Century Tibet Issue. Symposium on Contemporary Tibetan Studies. Taipeh, Taiwan. Retrieved from www.case.edu/affil/**tibet/tibetan**Nomads/documents/Taiwan_art.doc.

Cruikshank, J. (1985). The gravel magnet: Some social impacts of the Alaska highway on Yukon Indians. In K. Coates (Ed.), *The Alaska highway: Papers of the 40th anniversary symposium* (pp. 172–187). Vancouver: University of British Columbia Press.

Dreyer, J. T. (2003). Economic development in Tibet under the People's Republic of China. *Journal of Contemporary China, 12*(36), 411–430.

Fearnside, P. (1987). Deforestation and international economic development projects in Brazilian Amazonia. *Conservation Biology, 1(3)*, 214–221. Retrieved from http://www.jstor.org.proxy.library.cornell.edu/stable/2385877.

Gautam, H. (2014, June 13). Foreign employment lifts Dalit families out of poverty, Republica. Retrieved from http://www.myrepublica.com/portal/index.php/thweek/slc/ads/ncell.swf?action=news_details&news_id=78889.

Government of Nepal, Ministry of Finance (GoN/MoF). (2009). *Budget speech of fiscal year 2009–2010*. Retrieved from http://www.mof.gov.np/.

Gurung, H., Gurung, Y., & Chidi, C. L. (2006). *Nepal atlas of ethnic and caste groups*. Kathmandu: National Federation for Development of Indigenous Nationalities.

Gutschow, N. (1999). *Pilgrimage and space: The definatory purpose of pilgrimage routes: Case studies from Bhaktapur, Kag and Muktinath (Nepal) and Varanasi*. Contribution for the Proceedings of the Conference "Pilgrimage and Complexity" at Indira Gandhi National Centre for the Arts in New Delhi, January 5–9, 1999. Retrieved from http://www.colorado.edu/Conferences/pilgrimage/papers99/Gutschow.html.

Huber, T. (1999). *The cult of pure crystal mountain: Popular pilgrimage and visionary landscape in Southeast Tibet*. New York: Oxford University Press.

International Labour Organization (ILO). (2005). *HIV/AIDS and work, using the ILO code of practice on HIV/AIDS and the World of work: Guidelines for the transport sector, 2*. Retrieved from http://www.ilo.org/public/english/dialogue/sector/papers/transport/transp-hivguidlines.pdf.

Keck, M. (1998). Planafloro in rondonia: The limits of leverage. In J. A. Fox (Ed.), *The struggle for accountability: The World Bank, NGOs, and grassroots movements* (pp. 181–218). Cambridge, MA: MIT Press.

Kreutzman, H. (2004). Pastoral practices and their transformation in the Northwest Karakoram. *Nomadic Peoples, 8*(2), 54–88.

Laurence, R. (1999). *The roads of roman Italy: Mobility and cultural change*. London: Routledge.

Liechty, M. (2003). *Suitably modern: Making middle-class culture in a new consumer society*. Princeton, NJ: Princeton University Press.

Lutzenberger, J. (1985). The World Bank's Polonoroeste project: A social and environmental catastrophe. *Ecologist, 15*(1–2), 69–72.

Messerschmidt, D. (1992). *Muktinath: Himalayan pilgrimage, a cultural and historical guide*. Kathmandu: Sahaugi Press.

Murphy, M., Oli, K., Gorzula, S., & International Institute for Sustainable Development. (2005). *Conservation in conflict: The impact of the Maoist-government conflict on conservation and biodiversity in Nepal*. Winnipeg: Manitoba International Institute for Sustainable Development. Retrieved from http://www.iisd.org/pdf/2005/security_conservation_nepal.pdf.

National Trust for Nature Conservation (NTNC). (2008). *Sustainable development plan of mustang (2008–2013)*. Kathmandu, Nepal: NTNC/GoN/UNEP. Retrieved from http://www.rrcap. unep.org/nsds/uploadedfiles/file/sa/np/mnmt/document/sd_masterplan_Mustang.pdf.

Needham, N. (1971). *Science and civilisation in China. Volume 4, Part 3: Civil engineering and nantics*. Cambridge: Cambridge University Press.

Peet, R. (2003). *Unholy trinity: The IMF, World Bank, and WTO*. New York: Zed Books Ltd..

Price, D. P. (1989). *Before the bulldozer: The Nambiquara Indians and the World Bank*. Maryland: Seven Locks Press.

Salick, J., Yongping, Y., & Amend, A. (2005). Tibetan land use and change near Khawa Karpo, eastern Himalayas. *Economic Botany, 59*(4), 312–325.

Sax, W. S. (1990). Village daughter, village goddess: Residence, gender, and politics in a Himalayan pilgrimage. *Ethnologist, 17*(3), 491–512. Retrieved from http://www.jstor.org/stable/644858.

Shah, N. (2007). *Transport sector performance and impact indicators-A Nepalese case study*. World Bank Report. Retrieved from http://www.worldbank.org/transport/transportresults/program/sl-04-05/annex16.pdf.

Tuchman, B. W. (2001). *Stilwell and the American Experience in China, 1911–45*. New York: Grove Press.

Turner, M. (2010, April 1). *Coburn describes travels in Nepal. The Dartmouth*. Retrieved from http://thedartmouth.com/2010/04/01/news/shangrila.

Two new cable cars on the outskirts of the Kathmandu Valley. (2009, December 1). *ktm2day.com*. Retrieved from http://www.ktm2day.com/2009/12/01/two-new-cable-cars-on-the-outskirts-of-kathmandu-valley/.

United Nations Development Program (UNDP). (2006). *Developing rural transport and infrastructure. Millennium development goals needs assessment for Nepal*. Retrieved from http://www.undp.org.np/publication/html/mdg_NAN/Chapter_9.pdf.

Whelpton, J. (2005). *A history of Nepal*. Cambridge: Cambridge University Press.

World Bank. (2006). *Infrastructure: Lessons from the last two decades of World Bank Engagement*. Discussion Paper. Retrieved from http://www-wds.worldbank.org.

Part IV
The Way Forward

Chapter 6
The Future of Himalayan Mobilities

Abstract The 10 year People's War in Nepal (1986–2006) had a profound impact on infrastructure. Acts of sabotage which destroyed existing roads and bridges as well as withdrawal of foreign investment curtailed most construction projects. With formation of a new government in the wake of the war reconstruction and new development is on the rise-one of the key focuses is on connecting all 75-district headquarters by road. A renewed push for better north-south connectivity is integrated in a plan for regional trade between India, Nepal, and China. The precarious geo-political underpinnings of Nepal's geographical position between India and China makes road building a very complex multi-scalar proposition. The complexity of this challenge speaks to the importance of generating interdisciplinary research and development planning-which will necessitate cooperation and collaboration. The so-called Trans Himalayan Highway (a.k.a. Rasuwagadhi-Galchi-Raxaul Highway) is the shortest overland route from Tibet to India. As such it features prominently in many sub regional, regional and potentially international trade and commerce proposals for connecting China, India, South Asia and Europe. The Chinese financed Rasuwagadhi-Syaphrubesi section of this highway has been under intense scrutiny—it is being groomed to become the main access point for China-Nepal-India overland trade. Officially opened in December 2014 the Rasuwagadhi border has undergone huge changes recently.

First, from the forced closing of the Friendship Highway following a landslide in August 2014, which rerouted traffic to Rasuwagadhi and a temporary early opening in August 2014. Second, by the 2015 Gorkha Earthquake and the after math of destruction of both the road and the border facilities on either side. Third, the promulgation of the new Nepali constitution (September 2015) and the resulting embargo on Nepal's southern border leading to fuel supplies from China through Rasuwaghadi. And finally the extension of the Qinghai-Tibet Railway to Shigatse, Tibet with the intended continuation to the Tibet border town of Kyirong, just north of Rasuwaghadi and talks of extending the rail to Kathmandu and possibly Lumbini on Nepal's southern border. These changes are embedded in a matrix of discourse concerning trilateral trade, geopolitical influence, and national security between China, Nepal, and India. Ultimately Nepal is in a position to benefit from the aid and influence coming from both Beijing and Delhi, but a persistent and crippling

© The Author(s) 2017
R.E. Beazley, J.P. Lassoie, *Himalayan Mobilities*, SpringerBriefs
in Environmental Science, DOI 10.1007/978-3-319-55757-1_6

tradition of corruption and nepotism embedded in the chronically unstable Nepali government threatens to undermine any concrete benefit for Nepali citizens and keep the country locked into a decades old pattern of failed development programs and infrastructure stagnation. While the future holds many concrete substantial opportunities to develop new Himalayan Mobilities such as China's One Belt One Road Initiative among many others, the way in which these opportunities play out is dependent on Nepal's ability to curb its endemic political instability, corruption, and nepotism, and the greater economic and geopolitical developments of China, India, Asia, and Europe.

Keywords Trilateral Trade • China • India • South Asia • Europe • Sub Regional • Regional • Border • Southern Blockade • Nepali Constitution • Corruption • Nepotism • Rasuwagadhi • Friendship Highway • Qinghai-Tibet Railway • Tibet • Shigatse • Kyirong • One Belt One Road Initiative

When, in 1963, I asked the Nepalese: 'When will the China Road be finished?' they answered: 'It will take a long time yet because there is a considerable section to be cut through rock, which will be difficult.' The Nepalese could have avoided this error of judgment which underestimated the ability of the Chinese; the latter acquired great experience during the fifteen years they spent constructing roads in Tibet. When I put the same question this year (1965), the reply was: 'Oh, the new road is getting on very fast! It is almost finished … jeeps use it every day.' (Maillart 1966: 143)

China's road-building has surpassed even its own ambitious plans. The Ministry of Transport said Tuesday that construction of twelve national highways has been completed 13 years ahead of schedule. Another eight highways in western China are almost complete, as well, it said. The massive buildup of China's highways will soon leave the U.S., the originator of the national highway system, in the dust. Expressways in China now total 74,000 kilometers, or 46,000 miles, the ministry says—just a thousand miles short of the U.S. interstate system, according to U.S. government data. China has said that by 2020, China hopes to have about 85,000 kilometers of national expressways—a target that it will likely reach before the date, since it has already built 90% of the total. (Yan 2011)

During the 10-year, People's War in Nepal (1986–2006) development projects came to a standstill due to political instability, acts of sabotage, and withdrawal of foreign investments. After forming a new government in the wake of this conflict development projects are now on the rise, with road connection to all 75-district headquarters being one of the government's key goals. Several of the new roads under construction will connect to the Chinese border, opening up the potential for major trade between India, Nepal, and China. However, who will benefit the most from this trade and how the benefits will be distributed is hard to predict due to the intricate coupled connections between environmental, socioeconomic, and sociocultural spheres. The complexity of this problem speaks to the importance of generating interdisciplinary research and development projects, which will necessitate cooperation and collaboration. Owing to the inter-relationships between these three spheres, failing to address one of them will likely result in unintended consequences that will have negative affects on villages adjacent to these new commerce corridors.

This chapter uses the findings and conclusions discussed in Chaps. 2–5 as a framework to discuss emerging trends in Himalayan mobilities, and the effects of three recent seminal events, the major earthquake on April 25, 2015 and its after shocks, the promulgation of the new Nepali constitution (September 20, 2015), and the southern blockade (September 23, 2015–February 8, 2016) (Bhattacherjee 2016). Finally we offer suggestions and recommendations for future research on road development issues in Nepal and possibly elsewhere across the developing world.

6.1   New Himalayan Mobilities: Case Studies

The following section is based on the senior author's most recent research on road impacts and gendered mobility in the Trishuli River Valley of Rasuwa District in Central Nepal from April 2014–August 2015 (Fig. 6.1).

6.1.1   Enhancing North South Connectivity

Nepal has nine official border crossings with China (WB 2005) (Fig. 6.1). It is no secret that Nepal and China have been considering which of these border crossing would best be developed as transportation corridors in the future, as evidenced by the North South Transport Corridor Project. The North South Transport Corridor Project seeks to:

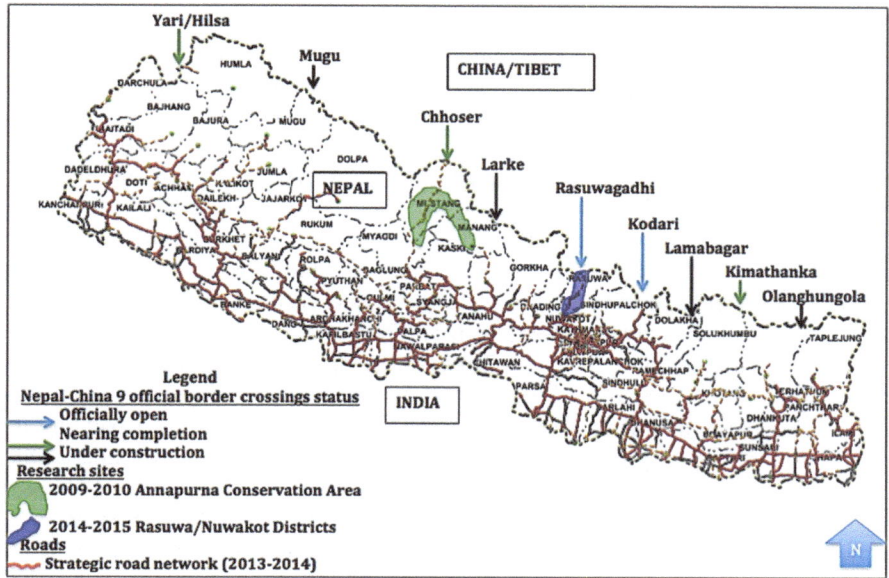

Fig. 6.1 Research sites and Nepal-China border crossings (adapted from GoN/DoR Strategic Road Network 2013–2014)

...Improve Nepal's north-south transport corridors, aimed at enhancing the accessibility of remote hill districts and at improved market integration and trade facilitation. In this context, the concept of north-south transport corridors involves three distinct elements: (i) development of Feeder Roads to link unconnected District Headquarters with Nepal's Strategic Road Network (SRN); (ii) construction of additional cross-border links with China; and (iii) construction of a new direct link between Kathmandu and the Terai, also known as the "Fast Track". (WB 2005: 5)

In addition to the Arniko Highway completed in the 1960s (see Sect. 1.5) it has only been recently that one of the other road links, the so called Trans Himalayan Highway (Bohara 2010) has been completed to the Chinese border at Rasuwaghadi (Rasuwa District) and another at Chhoser in Upper Mustang is nearing completion (see Fig. 6.1). A fourth border crossing at Yari in the western district of Humla already has a Chinese road link to the Tibetan border and the beginning of a Nepalese road is slowly making its way south from the border toward Simikot (see Fig. 6.1).

6.1.2   The Trans Himalayan Highway

The speed of China's buildup is also impressive, especially in recent years. It has built the majority of its expressways in the past decade and built 33,000 kilometers in the past five years. The U.S. Interstate system, in contrast, was built over more than three decades, starting from 1958 and lasting until 1991. (Yan 2011)

The Trans Himalayan Highway runs through the Trishuli River Valley in Rasuwa District northwest of Kathmandu (Fig. 6.2). It is both a transportation corridor and a geo-spatial concept linking China to India overland through Nepal. It is not a new route or concept as Nepal figured prominently during various periods of Trans Himalayan overland trade between its northern and southern neighbours even before Nepal became a unified nation in the late eighteenth century.

There were various trade routes that connected South Asia to Tibet and China through the Nepalese Himalaya with further extensions to other regional trade networks including the Silk Road network (Gleba et al. 2016; von der Heide 2012), the Southwest Silk Road network (Yang 2009), and the Ancient Tea-Horse Road (Sigley 2012; Fuquan 2004) (Fig. 6.3). Of these trans Himalayan routes the two closest links to Kathmandu, through the border towns at Kodari and Rasuwagadhi, were the most popular and heavily used (see Fig. 6.1). The Chinese in the 1970s helped build Nepal's first motorable road to the Tibet border approximating the Kodari trade route. In 2012 the second road to the border was completed, again with Chinese help, along the Syaphrubesi-Rasuwagadhi route (Fig. 6.2). This new road and its connection to Nepal's strategic road network features prominently in China's One Belt One Road (OBOR) economic connectivity initiative (formerly called China's "Silk Road Economic Belt and Twenty-First-Century Maritime Silk Road") announced in 2013 by President Xi Jinping - it is the shortest overland route from the Chinese border in Tibet to the Indian border (Fig. 6.4).

Fig. 6.2 The Kathmandu-Syaphrubesi-Somdang road and the newly constructed Chinese financed Syaphrubesi-Rasuwagadhi road (adapted from Google Maps)

Fig. 6.3 The ancient tea horse road and its connection to Nepal (Wikipedia)

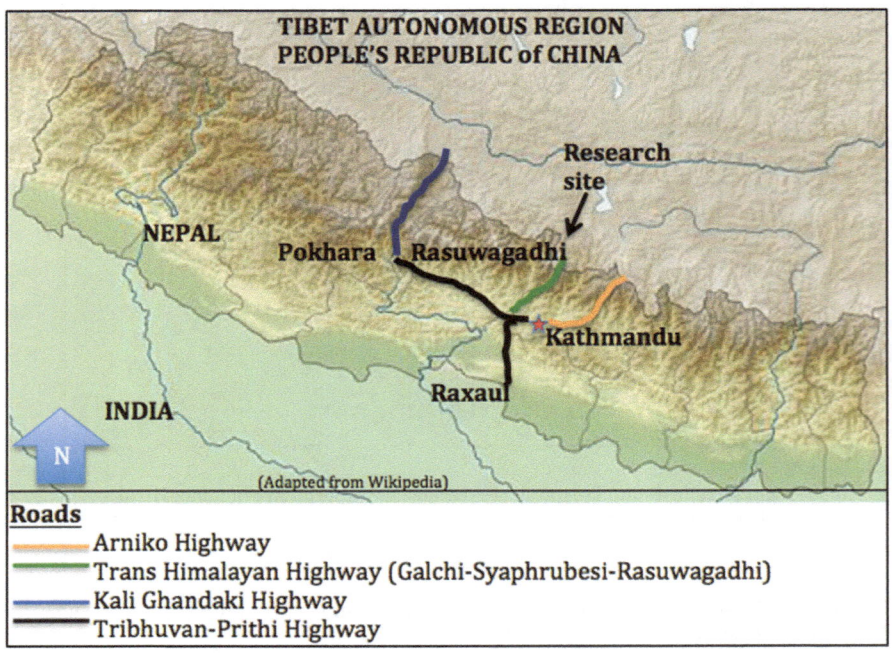

Fig. 6.4 The shortest land route between China and India—Rasuwagadhi to Raxaul with research site and other main north south roads

6.1.2.1 Historical Context

Stretching from the Kyirong (Tibetan *sKyid rong*) Valley translated as 'Happy Valley'[1] of the *Mang-yul*[2] *sbas yul* (*beyul*)[3] in the former Tibetan Kingdom of *Gung thang*[4] (thirteenth to fifteenth century AD) south through the valley of Shiva's trident (*Trishuli*)[5] lies an ancient mobility corridor that has witnessed a colorful cast of characters that date back in recorded history to the Tang Dynasty (618–907 AD) (Fig. 6.5). A stone inscription in present day Kyirong (a.k.a. Gyilong/Gyirong) County of the Tibet Autonomous Region (T.A.R.) dated 658 from the Tang mission to India bears witness to one of the first documented Sino-India diplomatic missions as well as to the reputation of this trans Himalayan

Fig. 6.5 Mangyul-Kyirong-Trishuli River power corridor and ancient trade route between Tibet, Nepal, and India (adapted from Wikipedia)

---

[1] *sKyid rong* Tibetan pronounced Kyirong meaning "Happy Valley".

[2] *Mang-yul* is the Tibetan name of a valley *sbas yul* pronounced *beyul* (extending north from the present day Nepal-Tibet boundary at Rasuwagadhi) in the ancient Tibetan Kingdom of *Gung thang*.

[3] *sbas yul* Tibetan pronounced *beyul* meaning "hidden lands" refers to valley sanctuaries which according to legend were hidden away by the great Buddhist tantric adept Padmasambhava to be used in times of need. "The hidden land is both a refuge for meritorious individuals from all strata of Tibetan society during a time of moral and political degeneration, as well as a place of accomplishment for those who are spiritually inclined. It is a land where the yogi can spend extended time in retreat and where a *gter-ston* can reveal sacred treasures." Childs 1999: 128) *gter-ston* Tibetan pronounced terton refers to a treasure revealer of sacred texts hidden by Padmasambhava and his consort Yeshe Tsgoyal.

[4] *Gung thang* Kingdom was founded under Sakya rule in the thirteenth century C.E. by Bumdegon (1253–1280) lasted until 1620 when the King of Tsang conquered it.

[5] *Trishuli* Sanskrit meaning the Hindu God Shiva's trident.

corridor (Lanxing 1992). Owing to its relatively low over all elevation in comparison to other trans Himalayan corridors it was often the prime route of choice in the middle Himalaya for north south trans Himalayan movement. It was both highway and Internet as trade and pilgrimage transferred goods and information along its path. Cultural influences and technologies including papermaking, printing, Nepali Newar architecture, the beginning of the Tibetan written script, and the first and latter transmissions of Buddhism to Tibet travelled along this corridor. A colorful cast of characters were instrumental in carrying these and other cultural catalysts. Princess Bhrikuti, Padmasambhava, Sakya Pandita, Atisha, Milarepa, the Madman of Tsang (Tsangnyon Heruka), Malla Kings, Qing Empire Imperial troops, Jung Bahadur Rana's Gorkha royal assassins, the zombie slayers of Langtang, the Great Trigonometric Survey Pundits, Heinrich Herrar, Peter Aufschnaiter, and many other luminaries footprints are embedded in the Trans Himalayan Highway as emissaries of these cultural transmissions (Beazley 2016a).

> Over the centuries, Mangyul-Gungthang has been an important gateway between the north and the south of the Himalayas, traversed by the main route between Tibet and Nepal, which passed through the Kyirong valley and led to Kathmandu. This was, at times, also an important route between China and India. The area is known from documents and inscriptions that go back to the imperial period (the sixth to ninth centuries). On an overhanging rock not far from the ruins of the royal palace is a Chinese inscription left in 658 by Wang Xuance, a diplomat of the Tang imperial court who passed through on his way to and from India. (Diemberger 2007: 34)

The current changes in the Trishuli Valley are embedded in legacy effects (Liu et al. 2007) of a long history of international trade. This corridor was one of the main trade routes that framed what has become known as the Trans Himalayan Salt Trade (THST).

The THST was essential to the survival of food deficient villages in the Himalayas. The basic ingredients of THST were salt from Tibet traded for grains from southern Nepal and India (Fürer-Haimendorf 1975). The western Himalayan countries including Nepal, Sikkim, Bhutan, and Upper Assam had no significant internal sources of salt, but they did have access to grain, either within their own country or from India to the south. Grain is hard to grow in Tibet, where the climate and altitude limit the variety and amount of grain that will survive. Consequently, trade developed with the exchange of salt for grains, which included rice, barley, and wheat (Van Spengen 2000). Tibetan nomads collected salt in the many regional saline lakes in Tibet and transported it to villages, markets, and monasteries where it was further transported to numerous fairs and markets along Tibet's Himalayan border. Traders living in the Himalayan highlands south of the Tibetan border brought rice to the middle hills area of Nepal where exchange for grain and other products from the south took place. It thus provided an important livelihood strategy for people living in the marginal agricultural areas of the Himalayas (Van Spengen 2000). Fürer-Haimendorf (1975: 132) explains the importance of this activity in Nepal: "All along Nepal's northern borders there are zones of high altitude inhabited by populations of Tibetan speech and Buddhist faith, who derive part of their subsistence from trans-Himalayan commerce."

As one of the main transit corridors along the previously porous Tibetan border the Trishuli River Valley also witnessed all three of the wars with Tibet (1788, 1792, and 1855) (Bauer 2004; Regmi 1970).

The contemporary history of the valley in terms of a mobility corridor is closely linked to Nepal's burgeoning hydropower development (Lord 2014). The first road from Kathmandu into the Trishuli River Valley (Kathmandu-Trishuli road 68 km) was built from 1957–1963 for the express purpose of transporting materials to Nuwakot District headquarters at Bidur to build the Trishuli Hydro-electric Project (Basnyet 1989). The Nepal Army was given the task of extending the road 105 km north and west to access a zinc-lead mine at Somdang. By mid 1984 the Army had extended the road up the valley to connect to the Rasuwa District headquarters at Dhunche (Armington 1985). Again in 1989 (Shakya 2009) it was extended to Syaphrubesi and further eventually reaching the mine at Somdang in 1991 (Nepal Army 2010) (see Fig. 6.2). There is some speculation that the road was for defence reasons rather than economic as the mine never appeared to be producing anything significant (Campbell 1993) and now locals say it is closed. However rumors circulate that rubies were discovered, which is why the mine is shrouded in secrecy and innuendo complete with connections to the highest levels of government (Shrestha 2001).

Although these road extensions were specifically for the mine at the time, they subsequently were instrumental in constructing the power house for the Chilime Hydro Project at Syaphrubesi and a feeder road was built from the Samdong road down to the village of Chilime to build the intake and reservoir which supplies water to the powerhouse in Syaphrubesi (see Fig. 6.2). But until 2012 there still was no road connecting Syaphrubesi north to the Tibetan border at Rasuwaghadi, a distance of 16 km. This section of the road was financed and built by the Chinese with construction beginning in 2009 and finishing in 2012 (Murton et al. 2016) (see Fig. 6.2).

The Rasuwaghadi link was targeted for opening in the fall of 2014, but with customs and immigration facilities on both sides still under preparation the date was postponed. However, a landslide (August 2, 2014) that blocked the Arniko Highway, at that time the only motorable road connection with Tibet, focused renewed attention on the importance of opening the border at Rasuwaghadi. Stretching form the border at Rasuwaghadi down Trishuli River Valley to Galchi where it meets the main highway and then south to border near Raxaul (see Fig. 6.4) it is the shortest route (265 km) from the Chinese border to India, hence it is recognized as being of the upmost importance (Shrestha 2014). China's interest in exploiting this transport link is evidenced by the investment of US$ 200 million in upgrading a dry port on its side of the border at Kyirong, in addition to financing the road link in Nepal from Syaphrubesi to the border at Rasuwaghadi (Bohara 2010), which involved building numerous bridges including the big fly over bridge connecting the two borders and China's new monolithic customs and immigration facility on its side of the bridge (Fig. 6.6). In addition, they committed to finance building the dry port facilities on the Nepalese side, and have plans to extend the Chinese railway system from its current terminus at Shigatse to the border of Nepal at Rasuwaghadi via the Shigatse-Kyirong section (Shrestha 2014). Thinking further into the future, China stated that after reaching the Nepalese border the railway line could be extended south to Kathmandu and eventually further south to Lumbini, the birthplace of the Buddha, and the Indian border (Shrestha 2014).

Fig. 6.6 The new Chinese Customs and Immigration facility at Rasuwagadhi-the bridge crosses the Lend Kola River, which forms the border between China (*left*) and Nepal (*right*)

6.1.3   Arniko Highway Closing

In geophysical terms the Himalayas are a highly dynamic environment, constantly growing ever since the Indian plate collided with Asia. It is estimated that the Indian plate is moving northward at 15–50 mm per year, expanding the Himalayas both vertically and horizontally. Nepal is situated in the central Himalayas encompassing roughly one third of the entire mountain range. The combination of the tectonically active mountains, steep river valleys, thin soft soil cover, and intense rainfall during the relatively short summer monsoon period make landslides, floods, debris flows, and other types of mass wasting the most common natural disasters in Nepal. Road building is inherently challenging and costly in any mountain setting, but in the Nepal Himalayas it is even more so (Dahal et al. 2008).

During a week of intense monsoon rainfall in August 2014 one of the biggest landslides (500 m wide) in Nepal in recent history occurred in the Sun Koshi (a.k.a. Bhote Koshi) River Valley covering a 1 km section of the Arniko Highway (AH) downstream from the border at Kodari, at that time Nepal's only officially functioning connection to its northern border with China (Fig. 6.7). The landslide swept away three villages and several schools with an estimated death toll exceeding 130 persons. It blocked the river creating a lake and the potential for a landslide dam outburst flood. Two hydro plants were damaged and transmission lines were swept away leaving the area without power, requiring additional load shedding (power outage) hours in the capital Kathmandu. Power companies estimated losses of eight million Nepali

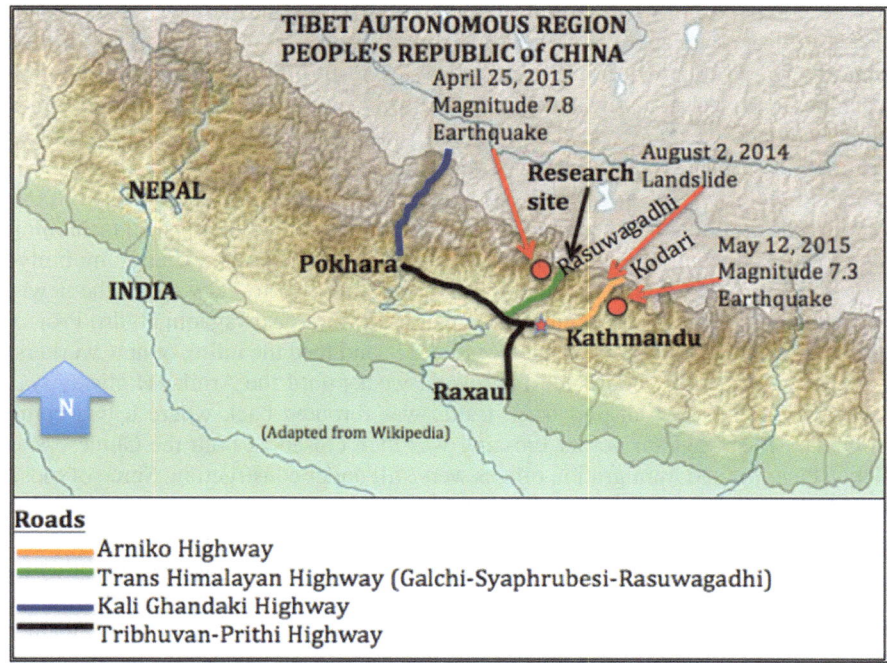

Fig. 6.7 Location of August 2, 2014 massive landslide on the Arniko Highway, the April 25, 2015 earthquake, the May 12, 2015 aftershock (earthquake) and major north south roads, in relation to author's research site

rupees per day (~US$ 80,000). Approximately 600 container trucks were stranded at the border with importers calculating their daily losses at 150 million Nepali rupees per day (~US$ 1.4 million). Most of the trucks were carrying food and goods for the upcoming Dasain festival in the fall raising speculation that there will be shortages and inflation for such goods as a result ("600 Containers Stranded" 2014). Villagers living above the landslide near the Chinese border (approximately 58,000 people) were cut off from the rest of Nepal resulting in lack of proper shelter, no access to relief aid, and food shortages. This lead to hoarding what goods were still available, producing a robust black market. Due to the scarcity of fuel local transport operators on the drivable upper section of the road increased the price of transportation to the point where locals were forced to walk (Manandhar 2014).

The solution to the problem was to temporarily open the Rasuwaghadi border crossing (see Fig. 6.7) ahead of schedule to handle the traffic diverted from the AH closure. This had a ripple effect that resulted in an influx of businessmen, entrepreneurs, laborers, truck drivers, government officials, and bank personnel all eager to take advantage of the economic benefits afforded by the temporary new border opening. Local labor organizers recruited porters from surrounding villages to transfer goods from Chinese to Nepali trucks. Both men and women arrived not only from the local area but from further south as the news travelled. This border area already had a large population of male migrants who came to work on the new Rasuwaghadi Hydro Project. Two labor camps housed approximately 200 Nepali and 150 Chinese laborers. Diesel transport trucks, four-wheel drive suvs, motorcycles, earthmoving machines, excavators, dozers, and dump trucks mixed with the dust and diesel fumes to create a landscape embedded with socially constructed symbols of hyper masculine mobility. Emergent in these hyper masculine mobility streams evolved stories of gendered mobility. Arriving female laborers became 'space invaders' (Puwar 2004) in what has traditionally been men's gendered work and mobility space, fashioning their own and complimentary mobility streams (or following Uteng 'mobility-scapes' 2011: 7). With this new economic activity women found work in the newly opened banks, in the hydro project labor camps, in the Rasuwagadhi Hydro Project office and many opened new businesses to house and feed the influx of new workers. This frenzy of activity continued for several weeks until the Arniko Highway was rehabilitated and the transport truck traffic was rerouted back where it had come from. The Rasuwaghadi border crossing was then closed as both the Chinese and Nepali customs and immigration offices were still under construction. Some of those who came up the road to unload the trucks headed back home, while others decided to stay in anticipation of more work in the future (Fig. 6.8).

This highlights the extremely political and context specific nature of mobility as well as how mobilities are nested in dynamic interconnected linkages within social and ecological systems. In this case one mobility corridor closing due to a landslide created economic hardships for the people living in the landslide area but opened up economic opportunities for people living in another area and affected gendered mobility patterns.

Fig. 6.8 Nepali women laborers unloading a Chinese truck

6.1.4   Hydro Project Built Roads

In the midst of Nepal's political instability following the People's War a new form of Himalayan mobility has emerged. This new mobility elucidates the coupled social ecological system of mobilities. In the Trishuli River Valley of Rasuwa District local roads are being built much faster than in many other parts of the country. The coupling of roads and hydropower development is event here. Roads are a prerequisite to bring the necessary materials, equipment, and labor for hydro project construction. The new Chinese built Syaphrubesi-Rasuwagadhi road was touted as a means to revive an old trade route, however since its completion the new the Rasuwagadhi Hydro Project, has begun near the border. This project would not have been feasible without the new road.

In the Trishuli Valley and adjacent watersheds there are currently four hydro projects operating, four under construction, and an additional 17 at some stage of the licensing process making a total of 25 projects (Fig. 6.9). As a result there is a spurt of road building extending up the west side of the valley from Betrawati through the towns of Shanti Bazar, Simle, and Mailung, to the current terminus at another hydro project site at Hakubesi, only 16 km south of Syaphrubesi (Fig. 6.9). The Nepalese Army is slated to complete the last remaining 16 km section of the road. The plan is for the new trade link from Rasuwagadhi to connect with the hydro project/Nepal Army west side roads near Syaphrubesi rather than use the old road on the east side through the district headquarters at Dhunche. The justification for this new route is compelling as it is straighter and shorter and it would eliminate having to traverse the three chronic seasonal landslide zones on the Dhunche road at Grang, Thade, and Ramche (Fig. 6.9). It will be faster and presumably safer.

This means the current main route to the border, the old road through Dhunche on the east side of the river (Fig. 6.9), will be eclipsed. People interviewed in Dhunche are understandably concerned that they will loose much of their current business as the new road will completely by pass them on the other side of the river. However, there are also two hydro projects proposed within Langtang National Park (on the east side of the river), which if approved might finance upgrading the Dhunche road to better standards and bring a new wave of businessmen, laborers, and additional income to the east side of the valley.

Additional rumors circulate that the whole new road system and its connection to the national highway (Prithi Highway) at Galchi will be widened to 30 m upgrading it to international highway standards to facilitate the new trade/customs port at Rasuwagadhi and ensuing truck transport traffic (Fig. 6.9).

6.1.4.1   Hydro Project 'Affected People'

Communities living in what are determined to be "hydro project affected areas" have been successful in leveraging hydro projects for community development projects as compensation for living in affected areas (Lord 2016, 2014; Shrestha et al. 2016).

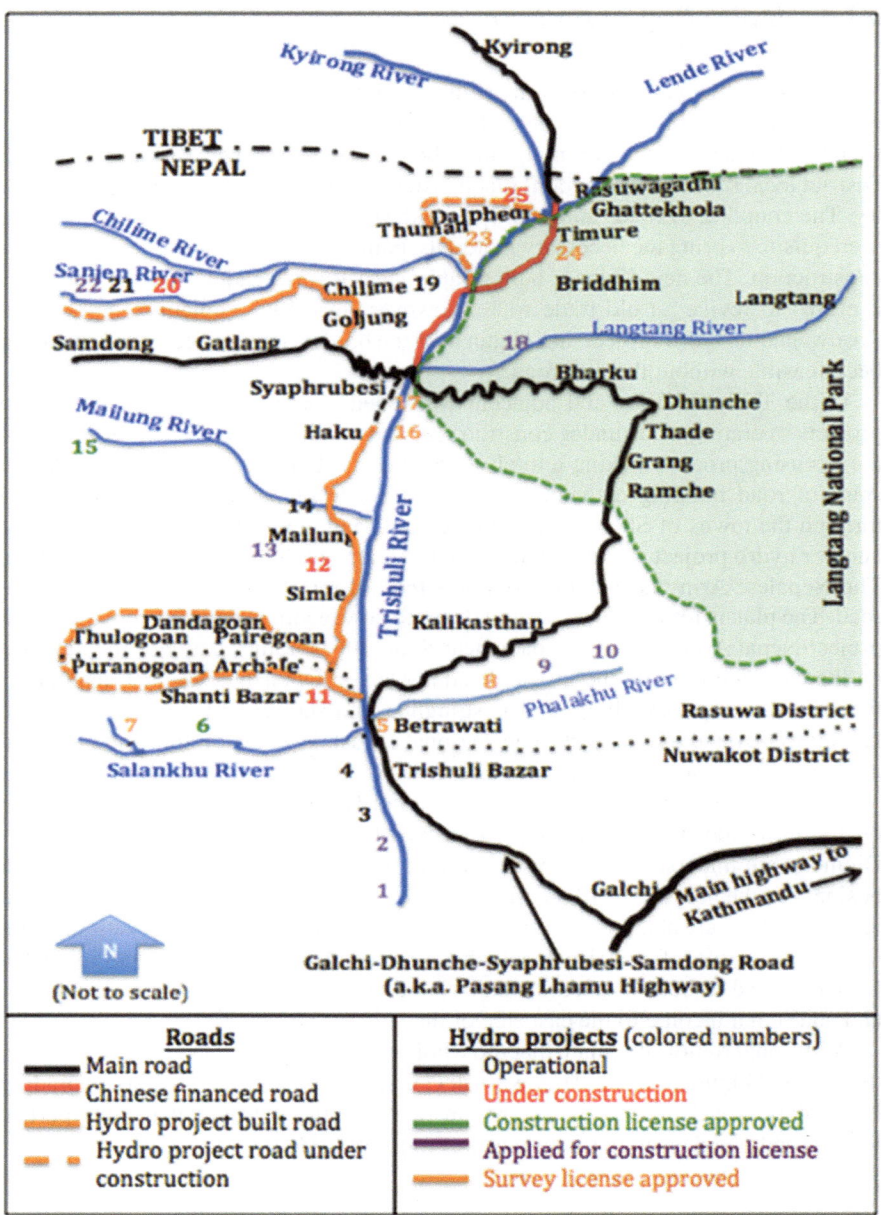

Fig. 6.9 Roads and hydro project sites Trishuli River Valley and adjacent watersheds (hydro project data source http://nitifoundation.org/hydro-map)

One of the most common requests is for a road to their community, in addition to school improvements and better water supplies. In many cases this has proven to be more efficacious than petitioning the Nepalese government through the traditional channels. It has the added advantage of accessing the readily available modern road building equipment hydro projects use to build their access roads. Consequently, it has the potential to build roads faster and arguably better.

On the west side of the Trishuli River near Betrawati a hydro project built loop road is under construction that will eventually connect the towns of Archale, Puranogaon, Thulogaon, Dandagoan, and Pairegaon (Fig. 6.10). Further north along the Syaphrubesi-Rasuwaghadi link new road construction to the village of Pajung is planned to continue up the Chilime River Valley as far as the scenic hot springs town at Tatopani (Fig. 6.10). Back in the main Trishuli Valley the Rasuwaghadi Hydro Project is building a road to Thuman, which may eventually be extended to Dalphedi and then connect to the main road again at Timure (Fig. 6.10). Just north of Timure the hydro project has built another road to the construction site of the new Armed Police Force compound (Fig. 6.10).

6.1.5   The April 2015 Earthquake and Its Aftermath

With this flurry of both hydro and road development in the Trishuli River Valley all eyes were focused on what was becoming the new model of development in Nepal (Murton et al. 2016). That is until April 25, 2015.

> On 25 April 2015 and over the next several weeks, a major series of displacements occurred ~15 km deep along the buried Main Himalayan Thrust without breaking the surface (1–3). The main shock of the Gorkha earthquake [magnitude (M) 7.8, U.S. Geological Survey (USGS); epicenter 28.147°N, 84.708°E] was followed by ~257 aftershocks of >M 3.0, including five ≥M 6.0 between 25 April and 10 June 2015.On 12 May, aM7.3aftershock struck~150 km ENE of the main shock. The largest earthquakes caused a wide swath of casualties and destruction in Nepal and adjacent India, China, and Bangladesh. Some mountain villages were shaken to complete destruction (4), buried by avalanches and landslides, or destroyed by powerful avalanche and landslide air blasts. The remote locations and blocked roads and rivers meant that ground crews could not immediately access many Himalayan valleys. (Kargel et al. 2015a: 1)

The "Gorkha earthquake sequence" (Collins and Jibson 2015: 1) affected eight million people, almost a quarter of Nepal's population (Adyan and Ulusay 2015) and created 2.8 million in need of humanitarian assistance (USAID September 2, 2015b). It is estimated that were close to 9000 casualties, 22,000 injuries, 605,254 houses destroyed, and 289,255 houses damaged (USAID December 23, 2015c) leaving approximately US$ 7 billion in economic losses (Hayes et al. 2015). Landslides were the main geophysical effect of the earthquake and its aftershocks. One group mapped more than 4000 landslides (Kargel et al. 2015a), greater than the total for the last 5 years (USAID June 25, 2015a) while estimates put the total number somewhere between 10,000–25,000 (Robb et al. 2015) with the total area affected by landslides in Nepal at 30,000 km^2 (areas impacted by landslides from the Nepal border extended

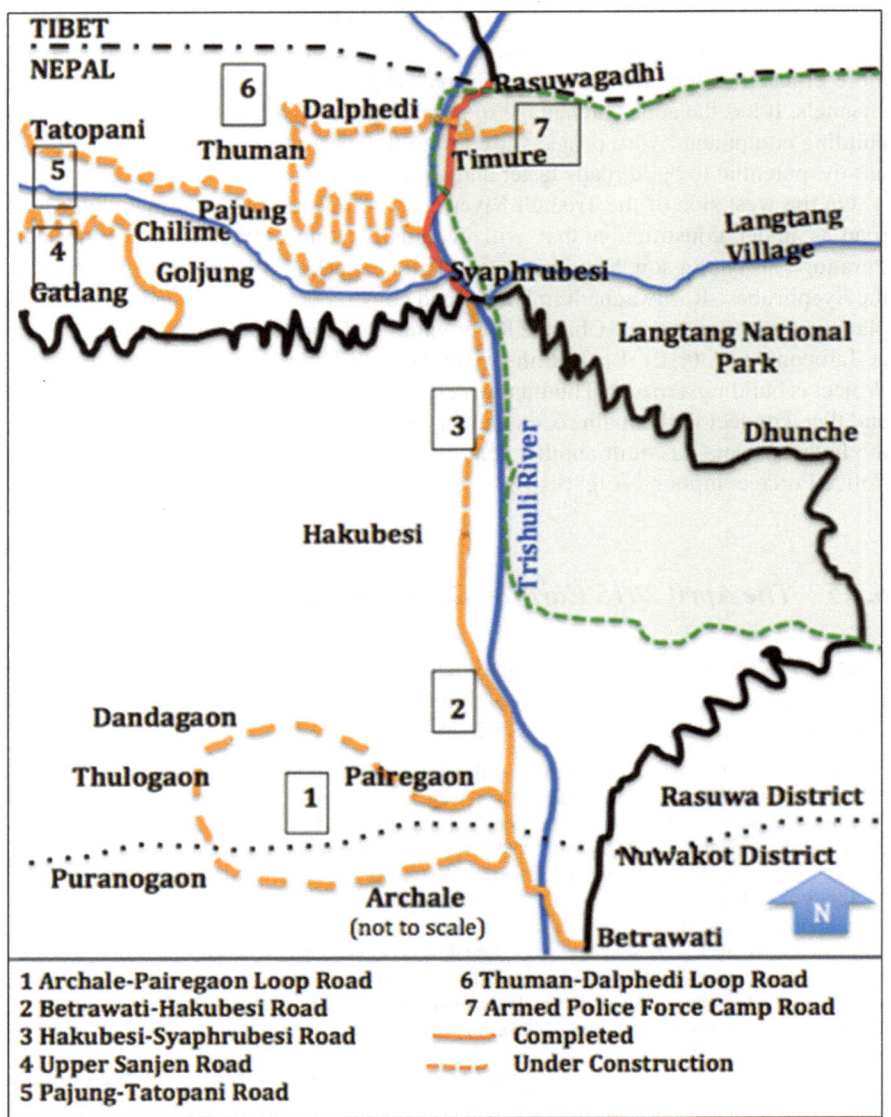

Fig. 6.10 Enlargement of hydro built roads map-Trishuli River Valley and adjacent watersheds

30 km into China which could increase this estimate by as much as 25%) (Collins and Jibson 2015). In addition to China the earthquake sequence caused casualties and damage in India and Bangladesh (Kargel et al. 2015a) affecting a total region approximately 500 by 200 km (Kargel et al. 2015b: 140) (see Fig. 6.7).

Rasuwa District is one of the 14 areas deemed hardest hit by the earthquake sequence and one of 12 areas of highest risk for landslides (Collins and Jibson 2015). Post earthquake large landslide numbers in Rasuwa were an estimated at 127 with a total area of 5,828,329 m² and a total debris volume of 2,914,165 m³ (GoN/ MoSTE 2015). However, total landslides of all sizes in Rasuwa was over 400 (Durham University 2015). The largest landslide of the Gorkha earthquake sequence occurred in the Langtang Valley, a tributary of the Trishuli Valley (Fig. 6.11). Triggered by the April 25 earthquake a debris avalanche of ice mixed with rock and soil dropped on the village of Langtang with a force equal to half of the atomic bomb dropped on Hiroshima creating winds in excess of 300 km/h (Kargel et al. 2015a, 2015b; Qiu 2015) that swept through the valley flattening trees and destroy- ing buildings in seven surrounding villages (ICIMOD 2015). The village of Langtang was completely covered destroying all buildings except one causing 350 casualties (Barry 2015) with a debris flow covering an area 900 m long by 400 m wide with an estimated volume of 2,000,000 m³ (Collins and Jibson 2015). Numerous other land- slides occurred in the Trishuli Valley with particularly large ones at Mailung, Gogane, Haku, and Rasuwagadhi (Fig. 6.11).

Hydro project personnel were evacuated from all the major project sites, in some cases by helicopter, as landslides damaged the road connections on both sides of the valley (Walker 2015). One assessment team counted more than 81 landslides along the Syaphrubesi-Rasuwagadhi road alone (GoN/MoSTE 2015: 33). What was once a model of new mobility and development suddenly became a broken and fractured landscape cut off from rescue and aid efforts coming from the south. The destruction

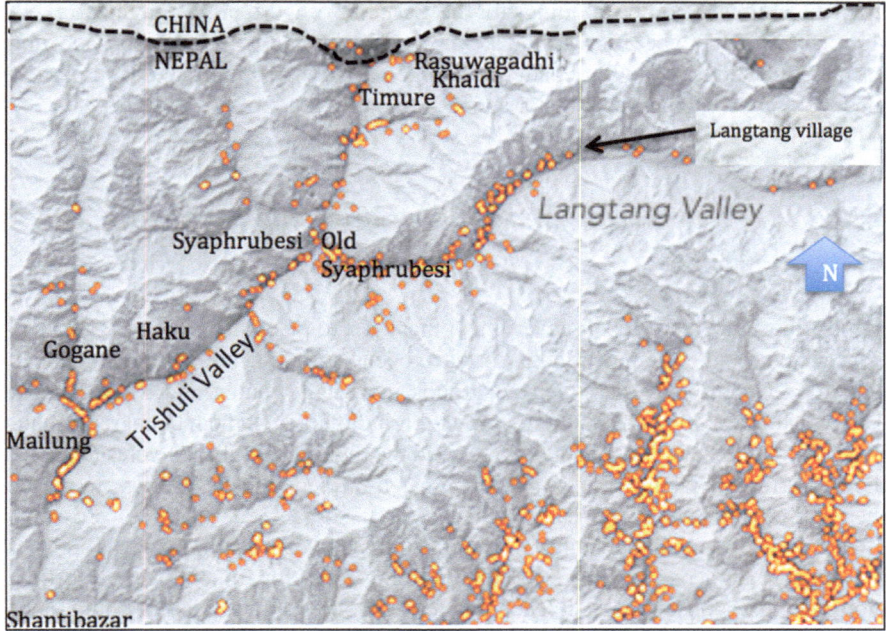

Fig. 6.11 Landslides after the April 2015 earthquake in the Langtang, Trishuli, and adjacent watersheds (adapted from NASA 2015)

in the Trishuli River Valley was mirrored in the Son Kosi/Bhoti Kosi River Valley to the east in Sindhupalchowk District. There the Arniko Highway and the border crossing at Kodari sustained major damage also cutting off aid and rescue efforts moving up from the south (Dangal 2015). Blocked old mobility patterns were abandoned as new ones were improvised to escape the death and destruction. In the south a mass exodus of more than one million people exited the capitol leaving sections of Kathmandu looking like a ghost town (Giri 2015). Yet while Kathmandu residents fled to relatively unaffected southern areas, refugees from the hard hit areas in the Himalayas arrived in Kathmandu seeking shelter and help. Numerous Internally Displaced People (IDP) camps sprung up around Kathmandu as well as in open spaces in the Trishuli River Valley and other valleys. Survivors from the Langtang Valley (at one point 125 households, almost 500 people) set up an IDP tent village within the Yellow Gumba Monastery (Phuntsok Choeling Monastery) walls near Swayambhu in Kathmandu (Sijapati et al. 2015).

In Rasuwa District an estimated 98% of the houses were either destroyed or uninhabitable leaving 2000 households (approximately 9000–10,000 people) in need of evacuation and temporary housing (Thapa 2015a). In the Trishuli River Valley the villagers of Timure near the border moved back up the mountain to their tradition home area of Khaidi (Fig. 6.12). Haku area villagers spread out through multiple IDP camps in Dhunche, Kalikasthan, Betrawati, and Kathmandu sometimes having to relocate several times as camps continued to mushroom in size (Fig. 6.12). Villagers from Mailung, Dandagaon, Thulogaon, and Gogane areas moved to IDP camps set up in the unfinished office buildings for the Trishuli 3A hydro project in Shanti Bazar (Fig. 6.12).[6] Everywhere in earthquake-affected areas as normal mobility corridors became blocked new mobility patterns emerged as survivors migrated from place to place looking for temporary shelter and aid (Fig. 6.12).

6.1.6   The New Nepali Constitution (2015) and the Southern Blockade

Amidst this chaos the Nepali government managed to push through a new constitution, which many speculate would not have been ratified, if not for the emergency situation. Traditionally marginalized groups including women and Madhesis and Tharus who live along the southern border, felt it did not give them equal representation (Haviland 2015). Nonetheless it received the required number of votes to pass in the Constituent Assembly and was promulgated on September 20, 2015. In protest the Madhesis immediately began a blockade of southern border road crossings, effectively stopping all vehicular movement. On the streets and in the press speculation pointed an accusing finger at Delhi blaming India for fomenting an unofficial blockade in response to Nepali leaders ignoring Delhi's suggestions for more inclusiveness in the new constitution (Pokharel 2015). Historically India has been the sole

---

[6] http://www.cccmnepal.org/DTM.

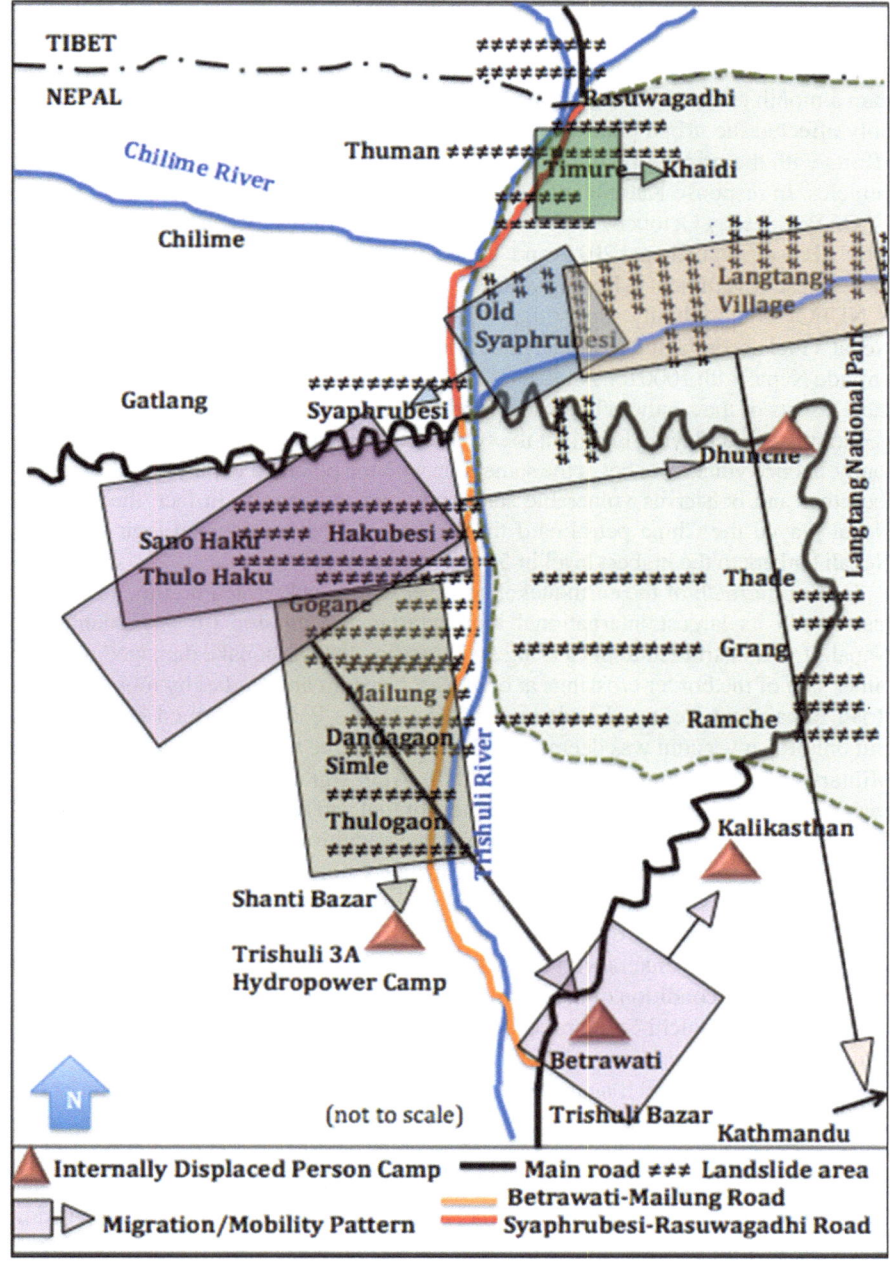

Fig. 6.12 Landslides and post earthquake migration/mobility patterns in the Trishuli Valley and adjacent watersheds

source of petrol supply to Nepal so the effects were immediately felt in Kathmandu and the rest of the country. The southern violence and demonstrations had effectively reduced the number of Nepali fuel trucks returning from India to a trickle for more than a month (Khanal 2015). As the blockade dragged on the restricted mobility not only affected the urban population but it also seriously hampered earthquake relief efforts with the price of petrol tripling and eventually no longer available for private vehicles. In response Kathmandu shifted its gaze north.

On Wednesday October 28, 2015 an eight member Nepali delegation signed a "historic" (Prasain and Khanal 2015) and "unprecedented" (Thapa 2015b) agreement with the Chinese government in Beijing whereby the China National United Fuel Corporation (CNUFC) would initially supply the Nepal Oil Corporation (NOC) with one third of Nepal's fuel needs "at international rates" (Pradhani 2015). In addition China agreed to provide Nepal with 1000 tons of fuel (Khanal 2015) as a grant to help relieve the immediate effects of the southern blockade. Media reports critiquing the China versus India geopolitical card playing highlight the two-sided coinage of Nepal's landlocked geography, at once vulnerable but at the same time with the potential to play one side against the other and bolster its vulnerable status into one of strength. In fact, the day after Nepal played the China petrol card the Indian Oil Company upped their supply to Nepali tankers to the highest level in 5 weeks (Khanal 2015).

In the aftermath of the earthquake, however China had wasted no time in sending historically its largest international humanitarian aid mission (to any country) to Nepal. It was instrumental in clearing and repairing the earthquake damaged roads on either side of the border crossings at both Rasuwagadhi and Kodari by mobilizing the Road Repair and Rescue detachment of the Chinese People's Armed Police Force. But only Rasuwagadhi was deemed ready for commerce again ("China Sends Record Military Personnel Numbers to Nepal" 2015). After signing the Chinese fuel deal, by early November 12 tankers with 80,000 liters of fuel had arrived in Kathmandu (Thapa 2015c) via the Rasuwagadhi border crossing and additional Chinese goods bound for Nepal had arrived in Kyirong ("Vehicles Carrying Chinese Imports to Nepal Arrive in Kyirong" 2015) as well as promises of liquid petroleum (cooking fuel). More over an additional 27 tankers were on their way back from Kyirong with another allotment, making a total of 71 tankers with Chinese fuel within 8 days (Samiti 2015a). Given the chronically poor condition of the Rasuwagadhi link to Kathmandu, the Pasang Lhamu Highway (a.k.a. Galchi-Syaphrubesi-Rasuwagadhi road), it is not surprising that the returning truck drivers immediately complained stating that it must be upgraded before they would travel it again (Rathaur 2015). Future plans to import more fuel on a commercial scale were peppered with talk about upgrading the Rasuwa highway route, a Chinese petrol pipeline from Kyirong to Nuwakot (Samiti 2015b) and an uninterrupted fuel supply once the railway from Shigatse, where the fuel supplies are stored, to Kyirong is completed. This has focused a renewed emphasis to open additional Nepal-China border crossings (Shrestha 2015c), and talk of a Trade and Transit Treaty giving Nepal access to Chinese sea ports facilitating third country trade ("Gov't Mulling Transit Treaty With China" 2015).

During the 28th Nepal China Border Customs Meeting (November 2–7, 2015) a number of comprehensive agreements were made to help ease commerce and trade between the two countries including improvement of infrastructure at both Kodari and Rasuwagadhi (see Fig. 6.1) as well as continued Chinese aid in constructing dry ports for both locations, a 20-point agreement to investigate border infrastructure effectiveness by a joint team, a renewed push to "operationalize" the seven other official border crossings, and permission for Nepali trucks, drivers, and traders to enter Chinese territory (Shrestha, P 2015c).

This is in keeping with a trend of increasing aid from China, which in 2015 made a fivefold jump from Rs. 2.4 billion (~US$ 23.9 million) annually to Rs. 13 billion (~US$ 129 million). On top of that Chinese President Xi Jimping on March 25 offered a Rs. 14.5 billion (US$ 144.1 million) economic aid portfolio earmarked to expand road construction and upgrade existing transport infrastructure (Shrestha, H 2015b). Following the earthquake China pledged an additional Rs. 76.4 billion (~US$ 759 million) in earthquake reconstruction aid ("China Pledged Largest Foreign Aid to Nepal" 2015) and a further Rs. 12.83 billion (~US$ 127.5 million) in grants for development of projects of mutual interest ("Foreign Aid Commitment Doubled" 2015). Along the Syaphrubesi-Rasuwagadhi road these "gifts of development" (Yeh 2013) have manifested not only in the road infrastructure mentioned previously (see Sect. 6.3.2) but in additional "gifts" including bags of rice, salt, Tibetan tea, tin roofing for local communities, and computers and monetary donations to upgrade local schools. Of course there is much speculation as to the 'why' and how China plans on manipulating this type of soft diplomacy termed both "development with Chinese characteristics" and "A Handshake Across the Himalayas" (Murton et al. 2016).

6.2 Future Regional Mobility Projects

Sandwiched between China and India, Nepal is situated to expand new Himalayan mobilities through various regional connectivity projects from both the north and south. From the north, China's Silk Road Economic Initiative also called the One Belt, One Road Initiative, was officially embraced when Nepal signed a four-point endorsement in December 2014 (Adhikari 2015). As part of this initiative Nepal's landlocked geography is geopolitically strategic for overland Sino-Indian trade routes as decades old Sino-Indian border disputes harbor old scars limiting overland transport through their shared boundaries, thus leaving Nepal with the potential to leverage a position of strength. As argued by Adhikari (2015: 15):

> Nepal is poised to take tremendous benefits from the silk route and resulting connectivity, which is associated with overall development of the country including massive infrastructure development. The opportunities include Silk Road USD 40 billion Fund; Asian Infrastructure and Investment Bank USD 50 billion; Enhanced connectivity resulting to increased Chinese aid and investment; Enhanced efficient and effective transit between India and China.

Many Nepali pundits see this as a win-win opportunity for Nepal bringing roads, railways, additional maritime route access, and multiple infrastructure investments, which can "significantly reduce the transaction costs and bring enormous economic benefits" transforming Nepal from a from a "landlocked to a land-linked state"

(Adhikari 2015: 9). In May 14, 2017 Nepal signed a MOU with China concerning the OBOR initiative. "'The MoU seeks to strengthen cooperation in connectivity sectors including transit transport, logistic systems, transport network and related infrastructure development such as railway, road, civil aviation, power grid, information and communication,' the statement said." (Giri 2015, para. 5)

On Nepal's southern border India has the third largest road network in the world (the United States is first followed by China)[7] and the third largest rail system (the United States is first followed by China).[8] In 2012 the National Highway Authority of India completed the Golden Quadrilateral, a massive 5846 km highway expansion program initiated in 2001 to connect India's four major cities of Delhi, Mumbai, Chennai, and Kolkata by a modern network of four and six lane express highways. This is only phase one of the National Highways Development Project.[9] On February 20, 2017 a statement by Suresh Prabhu, India's Ministry for Railways, announced that Kathmandu would soon be connected by rail to Delhi and Kolkata. This initiative is designed to increase mobility between the two countries by enhancing connectivity along their shared borders (India to link Kathmandu with Delhi, Kolkata by rail 2017). Previous to this in 2016 as a piece of the Bangladesh, Bhutan, India, Nepal (BBIN) Economic Corridor talk of a BBIN Rail Agreement modeled on the SAARC Regional Rail Agreement (RRA) has spawned additional sub-regional cooperation proposals on Connectivity and Transit and Sub-Regional Cooperation on Water Resources Management and Power/Hydropower (Bisht 2016).

There are numerous opportunities for Nepal to connect with India's highway and rail systems to further enhance future mobilities. One project recently completed (July 2015) is the B. P. Koirala Highway Project which links Kathmandu with cities in India via the south-eastern Terai border. The Sub Regional Transport Enhancement Project begun in 2010 seeks to enhance connectivity within Nepal to India and then though India to Bangladesh and Bhutan. The South Asia Subregional Economic Cooperation Road Connectivity Project (2013-ongoing) concentrates on improving connections with the Indian border in the east by improving and expanding Nepal's East-West Highway. Another project the Nepal India Trade and Transport Facilitation Project (2013-ongoing) focuses on reducing transportation time and costs by decreasing bottlenecks and modernizing border management (Rana and Karmacharya 2014). Finally the Trans-Asian Highway and the Trans-Asian Railway projects offer a myriad of potential new mobility corridors that would connect India through Nepal to China (Adhikari 2015).

While the new Himalayan mobilities evolving in the Trishuli River Valley suffered a temporary set back due to the earthquake it has resulted in a renewed push for more new mobility options in other parts of the country.

---

[7] http://www.roadtraffic-technology.com/features/featurethe-worlds-biggest-road-networks-4159235/.

[8] http://www.railway-technology.com/features/featurethe-worlds-longest-railway-networks-4180878/.

[9] http://www.roadtraffic-technology.com/projects/golden-quadrilateral-highway-network/.

6.3 Conclusions

6.3.1 Understanding the Environmental, Socioeconomic, and Sociocultural Couplings

In one sense, these north-south mobility corridors are not new phenomena. Many of the routes slated for roads, were former trade routes that flourished during various periods of the Trans Himalayan Trade. Kathmandu, for example, was an important entrêpot on a spur of the Silk Road (Rankin 2004). While it is often tempting to write about these valleys that are now connecting with motorable roads as remote and isolated, the history of Trans Himalayan Trade suggests that in comparison to other areas of Nepal that did not have major international trade routes these valleys were quite cosmopolitan, absorbing ideas and influences from many different cultural groups that traveled through them. Palaeoethnobotanical research suggests that the Kali Ghandaki Valley (Mustang District), for example, was an important international trade route as early as the first millennium BC (Knörzer 2000). Even in areas where there was not a major trade route enhanced mobility allowed some access to alternative livelihood strategies that others did not have. For example, the Manangis (indigenous people of Manang District) phenomenal rise in the business world over several centuries was due to trade and travel privileges afforded them by the Royal Government of Nepal that other groups did not have. Eventually these privileges helped Manangis extend trade networks internationally and in so doing bring not only economic capital back to their villages, but cultural influences as well (Rogers 2004).

Nonetheless, motorized vehicles bring changes and influences at a much faster rate and in greater volume. In the environmental sphere motorized vehicles bring air, noise, light, and water pollution, which has an affect on human health and wellbeing. However, better road connectivity provides easier and faster access to health, education, and social institutions, which can help promote better health conditions and socioeconomic wellbeing for communities. While these positive social benefits often accompany roads, there are other negative sociocultural impacts that have been well documented, such as the increase in the commercial sex trade, human trafficking, and the spread of so called "highway diseases" including HIV/AIDS (ADB 2008; Brushett and Osika 2005; UNDP 2006). Mobility facilitated by expanding rural road networks in the Himalayas is an intricately woven coupled social and ecological system. As such the effect of widening and hardening the earth surface to create a motorable road has a ripple effect that extends far beyond the immediate environmental construction site, which is often the most immediately recognizable impact. Our empirical research along the Kali Ghandaki and Marsyangdi Highways showed that roads have a zone of influence greater than 70 km beyond the immediate road surface (Fig. 6.13). These influences/impacts are not just limited to one sphere but include environmental, socioeconomic, and sociocultural spheres, which are multi-scalar (Beazley 2013).

Due to the complex connectivity inherent in the social and ecological system of Himalayan mobilities predicting the outcome of enhanced mobility projects such as

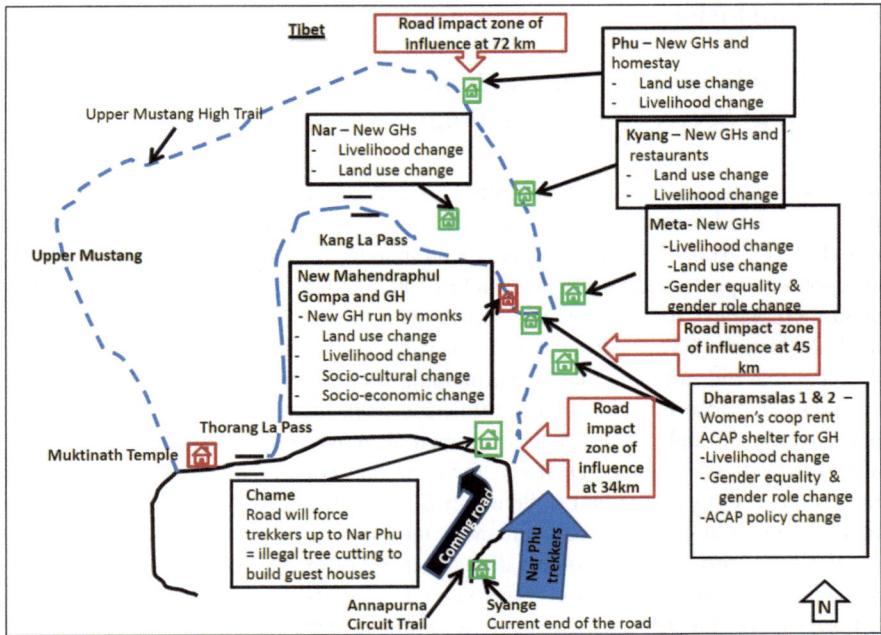

Fig. 6.13 Road impact zone of influence (Beazley 2013)

expanding mountain road networks is speculative at best and there are often unforeseen and unintended consequences. Different scales of mobilities are often nested in such a way that the linkages are not obvious and often have a delayed indirect impact effect as was seen in the case studies above (see Sect. 6.3).

6.3.2 The Future

Nepal is located between the world's second longest highway and rail networks (China) and the third longest highway and rail networks (India).[10] It is tempting to say that Nepal is on the verge of realizing a wave of new Himalayan mobilities with China offering to upgrade transportation infrastructure, improve and expand northern border crossing facilities, extend the railway to the border and eventually through Nepal to the Indian border, and welcoming Nepal into the One Belt One Road Initiative. From the south numerous initiatives including the B. P. Koirala Highway Project, the Sub Regional Transport Enhancement Project, the South Asia Subregional Economic Cooperation Road Connectivity Project, the Nepal India

[10] http://www.roadtraffic-technology.com/features/featurethe-worlds-biggest-road-networks-4159235/; http://www.railway-technology.com/features/featurethe-worlds-longest-railway-networks-4180878/.

Trade and Transport Facilitation Project, and the Trans-Asian Highway and the Trans-Asian Railway projects all hold the promise of creating the numerous southern regional links necessary to integrate with expanding northern mobility corridors and with the existing and expanding east-west transport networks.

Notwithstanding these possible imagined futures Nepal's unfolding development is not guaranteed, as evidenced by the following (Shakya 2016: 1):

> In early January, the Norwegian company Statkraft pulled out of the 650 MW Tamakoshi III Hydropower project after failing to make any major breakthrough with the government even after a decade of attempting to do so. This major news item went relatively unnoticed. The company spent more than $20 million in trying to figure out how best to invest over $1.5 billion in Nepal.

Even with the government promises and rhetoric about the "new Nepal" following the promulgation of the long awaited new constitution skepticism born from decades of broken promises and dashed dreams of becoming *bikasit*[11] is pervasive and deeply embedded in the Nepali subconscious.

In June 2016 at the Kathmandu Power Summit meeting PM Oli said, "After passing the constitution, Nepal has begun a new phase of development, and we are open for business. Invest your money in Nepal's energy development with confidence. Your investment will be safe, and the government is ready to extend the required support for this." (Rai 2016) However, past experience has made many countries wary:

> Many foreign investors are attracted by Nepal's huge hydropower potential, but are discouraged by the problems of corruption and geopolitics. A Chinese investor who attended the summit said, on condition of anonymity: "Government authorities seek bribes for everything, and there are fears among big Chinese investors that even if they produce surplus energy for Nepal, India is not going to buy it." (Rai 2016)

While other investors complain about bureaucratic red tape.

> "Administrative procedures are too lengthy, and government authorities are transferred too frequently," said Guo Jing, deputy project manager of the Rasuwagadhi Hydroelectric Project. "We will feel encouraged if these procedures are simplified." (Rai 2016)

The much-touted new transport route through Rasuwagadhi which is supposed to bring economic prosperity to the region (Cowan 2013) is a perfect example of how government corruption greases the wheels of those in the right circles and hinders the very people that are supposed to benefit from such *bikas*.[12]

> Once the trucks enter Nepal, customs processing is even more cumbersome. Each truck passes through 15 security checkpoints of Nepal Army, Armed Police Force and Nepal Police before they reach Kathmandu, and transporters say they have to pay off officials. "It takes us ten hours to make the journey of five hours to Trisuli," one driver complained.
>
> There are also hassles with immigration and security checks. Truck cartels and local extortionists charge up to Rs 5,000 per vehicle to let them through. Chinese tourists and business also say they have to pay off Nepali police. (Shrestha 2015a)

---

[11] *bikasit* is the Nepali word for developed.

[12] *bikas* is the Nepali word for development.

As we have seen there are many opportunities for developing new Himalayan mobilities but Nepal's ability to realize these opportunities is seriously compromised by a combination of burdensome and inefficient bureaucracy, a legacy of unstable governments, endemic corruption at all levels, and what has been called "the stifling burden of hierarchy"[13] (Foreign Hand 2016). The combination of these elements subsume development projects in such a way that the goal of the project is secondary to how many benefits can be siphoned off by national, regional, and local officials as the project moves forward. To get a sense of how crippling this modus operandi can be all one needs to do is look at the unnerving condition of roads in the Kathmandu Valley (see Snellinger 2013 and read the comments also-the description is accurate for the senior author's 2009–2010 and 2014–2105 research periods in Kathmandu) or read the newspaper-there is usually an article every week about Kathmandu's road problems (e.g. Gurung 2016). Even before the earthquake the roads in Kathmandu appeared to be foreshadowing the future-as if indeed an earthquake had already struck despite a portfolio of projects initiated to improve them including the Kathmandu Valley Road Improvement Project (2011-ongoing), the Kathmandu Ring Road Expansion (2013-ongoing), The Kathmandu Valley Road Expansion Project (2013-ongoing), the Kathmandu-Hetauda Fast Track Project (2009-ongoing), and the Kathmandu-Kulekhani-Hetauda Tunnel Highway (2013-ongoing). The reason projects like these take so long to complete (if they ever are completed) is a combination of factors including difficulties in equitable land acquisition and compensation, corrupt contract procurement practices, nepotism between the government and contractors, unscrupulous contractor and sub contractor practices, and corruption, graft, and payoffs through all levels of government from the village level through district to national level.

> The whiff of corruption swirls around road construction projects. Conversations with people about the history of any particular road would always at some point lead to a discussion of theft, embezzlement, nepotism, and shady dealings of one kind or other. And if there was nothing obvious to complain of, there was still the assumption that something underhanded was going on. The roads delivered are never quite the roads that were promised, either in their material quality or in their planned routing. Private interests are commonly assumed to derail the process: politicians and construction companies take a cut; contracts and sub-contracts are circulated through friends and families; funds and materials are siphoned off by those who hold positions of responsibility. (Harvey and Knox 2015: 345–346)

The corruption described by Harvey and Knox (2015) from years of ethnographic road research in Peru is doubly true for road construction projects in Nepal. In such an environment it is not a surprise that many projects appear to be unfinished-in a state of protracted limbo-waiting for the next yearly change of government with new ministers whose favored contractors and budget infusions will get it back on 'track', at least for a few months. It is a pattern that is duplicated in other cities like Biratnagar and which many have come to accept as normal (Tiwari 2016).

---

[13] This phrase refers to the political system in Nepal in which the outer workings claim to be a democracy but the inner workings of each party are "tightly controlled by senior leaders who stifle any attempt at internal democracy." (Foreign Hand 2016).

Even when roads are completed the benefits to the average citizen are filtered through more corruption and graft when transportation syndicates/cartels take over control of sections of highway (see Sect. 2.5).

The exception to this *ad hoc* and plodding road development model is the hydro project and hydro project affected people model described in Sect. 6.1.4. In the Trishuli River Valley roads are being started and completed. Some are hydro project roads built to access projects sites, which at the same time improve mobility for local communities, others are roads built by the hydro project to provide better access directly to the local communities as part of their social responsibility and duty to address concerns of project affected people (see Fig. 6.10) (Lord 2016; Shrestha et al. 2016). Either way the roads are built much faster then waiting for the central government to act. Whether this will become the new model for road development in other parts of Nepal is hard to predict but other valleys with hydro projects such as the Upper Tama Koshi Hydro Project and recently revived Arun III Hydro Project are boasting new roads and improved mobility for local inhabitants also (Fig. 6.14).

Nonetheless, infrastructure development and in particular road projects seem to go hand in hand with corruption regardless of the country—as in South America described above by Harvey and Knox (2015: 345–346) and in China by Lin (2011: 151) who comments:

> …the roads seem to be the most common and most effective way to move. Moreover, they suggest themselves as a development tool for local leaders eager to bring the benefits of the market economy to remote areas under their jurisdiction. At the same time, road construc-

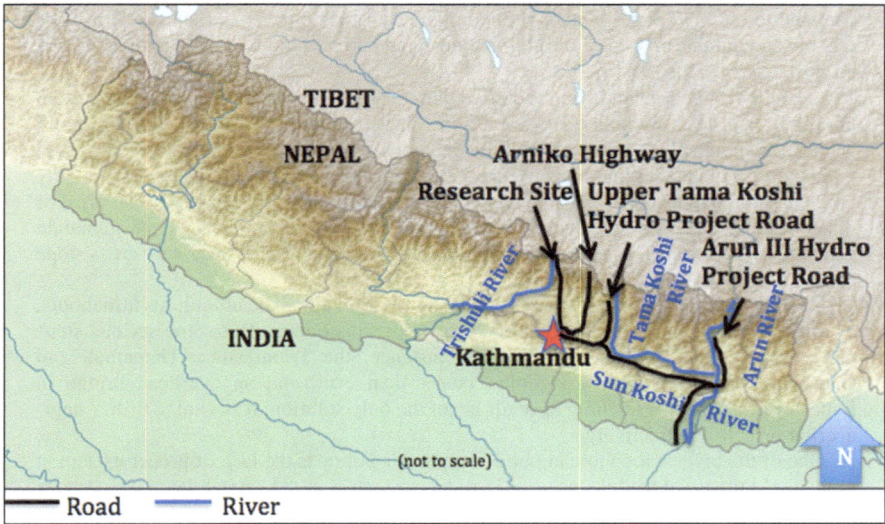

Fig. 6.14 Tama Koshi and Arun III hydro project roads in relation to author's research site (adapted from Wikipedia)

tion projects provide a breeding ground for corruption and abusive taxation, as well as environmental movements, protests from the base of the population, disagreements on land use and violent confrontations between local government and societal groups.

There is no lack of aid available for infrastructure development from neighboring countries such as India and China in addition to the multilateral aid institutions such as the World Bank, which in August 2016 approved US$ 55 million (Rs. 5.91 billion) to expand Nepal's Road Sector Development Project (RSDP) to aid in post earthquake reconstruction and upgrading ("WB Okays $55m Loan for Roads, Bridges" 2016). Nonetheless, if all the money does not reach its intended project then projects are often never completed or even started. A sobering report from first year Nepali undergraduate civil engineering student Sarin tells the story in detail (Pokhrel 2015).

A month of internship at Nepal Adarsha Nirman Company availed me with tons of exposure as my very first task as a civil engineer kicked off in a great style. It was an incredible experience to work as a trainee in the "Asphalt Road Maintenance Work, Naubise-Piplamod Road." During the course, I came across several facts about the actual scenario of the roads in Nepal.

First of all, it was painful to know that the Government of Nepal squander hefty sum in the name of maintenance work every year, but the task doesn't receive a clear finishing. Naubise road is the only way via which thousands of vehicles enter the valley every day. Nevertheless, the two-lane pavement is not sufficient enough for the smooth flow of traffic. Also, because it's narrow, the chances of accident soar up every now and then. Furthermore, the flexible pavement that covers the majority section of a road is designed in such a way that it can resist only 20 tons of load. Unfortunately the heavy weighted transport and the excessive loads carried by them together weigh almost 40 tons. As such the bearing capacity of the road gets deteriorated than its normal life span. Since the tenure of flexible pavement is less, we can build a rigid pavement throughout the Naubise road as the path is confined to only two lanes. Although the initial cost of rigid pavement is high, it is more reliable and durable. Wise use of reinforcement and correct ratio of concrete mix can give us even better results.

Secondly, while asphalt laying is carried out, it is necessary to keep away vehicles from the road for at least 24 hours. However, as we have an only access to enter the capital, vehicles are compelled to pass without the pitch being compacted properly. This results in faulty construction resulting in visible cracks on the newly constructed roads. It is not only in the case of Naubise road, most newly constructed roads are not designed properly. For illustration: In a heavy downpour, the recently constructed Shantinagar road in Kathmandu gets filled with water. There is no proper drainage system nor is the necessary cross slope maintained.

Yes, we are topographically weak and it's unfair to imagine the ubiquitous turnabouts, overhead bridges and cross over in every parts of the country. But some serious steps must be taken for the smooth flow. For instance, the Tripureshwar-Thapathali and Gaushala-Chabahil section are getting busier than ever, and an overhead bridge in Baneshwar is a must. Widening of roads is not the only solution; it is vital to utilize engineering principles effectively.

One of the biggest loophole in our government's policy is the lack of proper execution of plans. At times we read the news of introducing metros in the capital, but the news gets limited to papers and promises. Although the metro's construction is not a day's play, it won't happen ever if we don't even start the work. The saddest part is that most of the government officials are so busy to fend for themselves that they have even forgotten to utilize the budget sincerely. The interminable tales of corrupt engineers and contractors is not a new story in our nation.

> Whatever I have pointed out in this article is not even one third of the actual scenario of roads in Nepal. In a nutshell, instead of spending time in dividing the country into provinces, let's join our hands together to build a well-managed Nepal. [sic]

While it may seem easy to blame the government for all the problems in Nepal the fact remains that Nepal has a long history of failed development initiatives and squandered foreign aid and investment.

Fifteen years ago one of Nepal's most distinguished academics had already written a full book on the subject titled "Nepal's Failed Development" and keenly observed (Panday 1999: xiv):

> With the gap between rhetoric and action…, the development practices in the country may soon have consequences more hazardous than the ridicue they could invite from a Havel and a Freire for the underlying hypocrisy and ignorance….Nepal may not be the only country suffering from such pathology. But this is no relief for the majority of the country's citizens whose state of deprivation is matched only by the callous rapacity of the ruling class.

For a recent critique of this same subject (see Bell 2015a).

The legacy effects of decades of this default mode of operation by the government became blatantly obvious in its slow and ineffective response to the earthquakes and its questionable handling of the subsequent relief efforts mounted by both domestic and international agencies (Beazley 2016b). A full year after the April 25, 2015 earthquake there is still not one house that has been rebuilt by the Nepalese government despite the fact that billions of dollars (US$ 4.1 billion) have been donated and set aside for rebuilding (Rigby 2016). Entangled in this mess and part of the problem is the vacuum of local accountability left by the lack of local elections for more than a decade. As Bell (2015b) explains:

> In trying to understand what's going on, and what could be done about it, it's important to recognise that this emergency has, among other things, cruelly exposed the mess that Nepal was already living with.
>
> Many of the worst affected areas received little, or almost nothing, in terms of government or donor-driven support even before the earthquake. Health, education and other provisions (such as transport infrastructure) were minimal. In a mundane way, life was already pretty desperate there, making people more vulnerable to calamity.
>
> There has always been an incompetent administration and political class, which is obsessed with control of resources, but callous in its lack of urgency in providing for needy rural people.
>
> **What Didn't Work Before?**
>
> One major scheme of the past decade—designed to fund infrastructure construction—was called the Local Government and Community Development Programme. This delivered hundreds of millions of dollars through complex, donor-designed systems.
>
> In the absence of an elected local government, committees of unelected politicians called "All Party Mechanisms" misused vast sums intended for the poor, for the benefit of themselves and their cronies. There is once again talk of reviving All Party Mechanisms now.
>
> A few days before the earthquake, several donors withdrew from another major funding mechanism of the last decade called the Nepal Peace Trust Fund. It was never very clear how money from the NPTF was used, and the donors didn't explain their withdrawal. One of the NPTF's programmes was intended to provide compensation to victims of Nepal's conflict. In practice, many genuine victims received nothing, while district-level politicians and administrators gave the money to local supporters. This is worth remembering now when it comes to earthquake victims.

Whether a symptom, cause, or both the combined effects of decades of changing unstable governments and the ensuing 10 year Maoist War and its political aftermath have done little to change the situation despite hope that abolishing the monarchy and declaring Nepal a Federal Democratic Republic would bring a new era. From the beginning of the "democratic period" when Nepal became a constitutional monarchy (1990) to the dissolution of the monarch (2006) after the Maoist War peace agreement, a period of 16 years, Nepal had 16 different prime ministers. From 2008 when the Maoist party leader Pushpa Kamal Dahal was elected prime minister (for his first term) until the present (April 2017) Nepal has seen nine prime ministers in 8 years. In his first term as Prime Minister Dahal resigned after only 8 months due to a dispute over his firing of the army chief. In July 2016 then PM Oli stepped down anticipating a no confidence vote, he had been in office only 9 months (Bell 2014). Through a deal with the nation's largest party (Nepali Congress) Dahal, on August 3, 2016 once again became PM with the stipulation that he too step down after only 9 months.

> And that is the trouble with Nepal's politicians. They are more interested in squabbles over power than in solving problems. Poverty has sent millions of Nepalis abroad for work. The state has failed utterly to rebuild after an earthquake last year devastated mountain villages, killing 8,000 and leaving millions homeless. A new constitution has yet to be fairly implemented more than ten months after it was passed. ("An Ex-Revolutionary Becomes Prime Minister—Again" 2016).[33]

It is hard to believe that any country could initiate and follow through on any long-term infrastructure and development programs when the government is as unstable and changes hands as often as it does in Nepal. Nonetheless, the progress being made in the Trishuli Valley bodes well for the potential to realize new Himalayan mobilities. The 2015 earthquake and aftershocks have temporarily stopped this progress and for the time being efforts are directed at repairing what was damaged to pre earthquake conditions. It will take time to do this but the aftermath of the three big events of the 2015, the earthquakes, the new constitution, and the southern blockade have spurred a new enthusiasm for exploring and expanding new Himalayan mobilities. While the new Himalayan mobilities evolving in the Trishuli River Valley suffered a temporary set back due to the earthquake it has resulted in a renewed push for more new mobility options in other parts of the country.

6.4   Future Research

There are several areas that present themselves for future important research for Himalayan Mobilities within the context of post People's War and post 2015 earthquake Nepal.

First an emphasis must be placed on including the role of geologists and their knowledge of fault lines in the planning and execution of future road building. While this would increase the cost of projects in the short-term, in the long run it would pay for itself in terms of road repair and maintenance work. This has been proven time and again with the existing roads that have been built along fault lines and landslide prone areas that constantly need repair such as the Mahendra Highway (Regmi et al. 2014) and the Dhunche road (MacFarlane et al. 1992). Of course part of the problem is that road cuts if done poorly can create landslide prone areas. Incorporating best practice road building guidelines with fault line mapping is the best solution. Research on how this can be done within the present economic and political milieu of Nepal is essential.

Second using a coupled social and ecological system framework for road construction projects would naturally incorporate not only geologists but social scientists as well providing a better understanding of the geo-physical, socio-economic, and socio-cultural concerns important in road building (see Harvey and Knox 2015). The social cost of natural disasters is in sharp focus post earthquake and this knowledge needs to be incorporated into the rebuilding of not only roads but also all future infrastructure projects as well. In this regard gendered mobility should be a major component of future research.

Third, analyzing the outcomes of the two present border crossings at Kodari and Rasuwagadhi will be instrumental in the development of the future border crossing development presently in focus. Natural disasters like the Sun Koshi landslide and the earthquakes affected both border crossings and again a coupled systems approach to investing these incidents as well as others throughout Nepal and elsewhere (e.g., Chile earthquake 2016, Fukushima tsunami 2015, Haiti earthquake 2010, Sichuan earthquake 2008, Hurricane Katrina USA 2005, etc.) will help inform a more comprehensive development process.

Finally, post natural disaster Nepal provides fertile ground for further research into corruption, political instability, and inefficient bureaucracy as barriers to safe and equitable infrastructure development.

References

600 Containers Stranded. (2014, August 14). 600 containers stranded due to obstruction of highway. *Karobar Daily*. Retrieved from http://www.karobardaily.com/news/2014/08/600-containers-stranded-due-to-obstruction-of-highway.

Adhikari, R. (Ed.). (2015). *Silk route: Enhancing Nepal-China connectivity: Celebrating 60th year of the establishment of diplomatic relation between Nepal and China. Seminar on silk route, Kathmandu, march 23, 2015*. Kathmandu: Institute of Foreign Affair.

An Ex-revolutionary Becomes Prime Minister—Again. (2016, August 4). *The Economist*. Retrieved from http://www.economist.com/news/asia/21703439-appointment-unlikely-bring-much-needed-stability-country-ex-revolutionary.

Armington, S. (1985). *Trekking in the Nepal Himalaya* (4th ed.). South Yarra: Lonely Planet.

Asian Development Bank (ADB). (2008). *ADB, Roads, and HIV/AIDS: A resource book for the transport sector*. Retrieved from http://www.adb.org/Documents/Books/ADB-HIV-Toolkit/ADB-HIV-Toolkit.pdf.

Aydan, O., & Ulusay, R. (2015). *A quick report on the 2015 Gorkha (Nepal) earthquake and its geo-engineering aspects*. IAEG, 26pp

Barry, E. (2015, December 18). Nepal avalanche carried half the force of an Atomic Bomb. *The New York Times*. Retrieved from http://www.nytimes.com/2015/12/19/world/asia/nepal-avalanche-langtang.html.

Basnyet, S. (1989). *Micro–level environmental management: Observations on public and private responses in Kakani Panchayat, ICIMOD occasional paper no. 12.* Kathmandu: ICIMOD.

Bauer, K. (2004). *High frontiers.* New York: Columbia University Press.

Beazley, R. E. (2016a). *Divine Madmen, Zombie Slayers, Imperial Soldiers, and Royal Assassins: Footsteps along a himalayan power corridor.* Manuscript in preparation.

Beazley, R. E. (2016b). *Himalayan Trauma: Administrative thrombosis and citizens' response to 2015 Nepal earthquake.* Manuscript submitted for publication.

Beazley, R. E. (2013). *Impacts of expanding rural road networks on communities in the annapurna conservation area, Nepal.* M.S. Thesis, Department of Natural Resources, Cornell University, Ithaca, NY.

Bell, T. (2015a, March 22). *Nepal's failed development, Aljazeera.* Retrieved from http://www.aljazeera.com/indepth/opinion/2015/03/nepal-failed-development-150322052502920.html.

Bell, T. (2015b, May 25). *One month on, red tape hampers Nepal aid, Aljazeera, aljzeera.com.* Retrieved from http://www.aljazeera.com/indepth/opinion/2015/05/nepal-aid-earthquake-corruption-150524135716175.html.

Bell, T. (2014). *Kathmandu.* India: Random House India.

Bhattacherjee, K. (2016, February 8). Madhesi end Nepal blockade. *The Hindu.* Retrieved from http://www.thehindu.com/news/international/madhesis-end-nepal-blockade/article8210239.ece.

Bisht, R. (2016, January). Bangladesh, Bhutan, India, Nepal to discuss BBIN Rail Agreement. *Governance Today.* Retrieved from http://governancetoday.co.in/bangladesh-bhutan-india-nepal-to-discuss-bbin-rail-agreement/.

Bohara, A. (2010). *Prospects of a Trilateral Trans-Himalayan Economic Cooperation Agreement (THECA): China, Nepal, and India.* Retrieved from https://ejournals.unm.edu/index.php/nsc/article/view/223.

Brushett, S., & Osika, J. S. (2005). *Lessons learned to date from HIV/AIDS transport corridor projects.* World Bank Global HIV/AIDS Program. Retrieved from http://siteresources.worldbank.org/INTHIVAIDS/Resources/375798-1103037153392/Transport.pdf.

Campbell, B. (1993). The dynamics of cooperation: Households and economy in a tamang community of Nepal. Ph.D. dissertation. The School of Development Studies, University of East Anglia.

Childs, G. (1999). Refuge and revitalization: Hidden Himalayan sanctuaries (Sbas-yul) and the preservation of Tibet's Imperial Lineage. *Acta Orientalia, 60,* 126–158.

China Pledged Largest Foreign Aid to Nepal. (2015, August 8). *The Kathmandu Post.* Retrieved from http://kathmandupost.ekantipur.com/news/2015-08-24/china-pledged-largest-foreign-aid-to-nepal.html.

China Sends Record Military Personnel Numbers to Nepal. (2015, May 7). *China Daily.* (Xinhua). Retrieved from http://www.chinadaily.com.cn/china/2015-05/07/content_20652513.htm.

Collins, B., & Jibson, R. (2015). *Assessment of existing and potential landslide hazards resulting from the April 25, 2015 Gorkha, Nepal earthquake sequence (ver. 1.1, August 2015): U.S. Geological Survey Open-File Report 2015–1142* (p. 50). Retrieved from http://dx.doi.org/10.3133/ofr20151142.

Cowan, S. (2013). All change at rasuwa Garwi. *HIMALAYA: The Journal of the Association for Nepal and Himalayan Studies, 33,* 97–102.

Dahal, R., Hasegawa, S., Nonomura, A., Yamanaka, M., Dhakal, S., & Paudyal, P. (2008). Predictive modeling of rainfall-induced landslide hazard in the lesser himalaya of Nepal based on weights-of-evidence. *Geomorphology, 102,* 496–510.

Dangal, D. (2015, June 9). Post-quake, tatopani under existential threat. *Republica.* Retrieved from http://www.myrepublica.com/society/item/23656-post-quake-tatopani-under-existential-threat.html.

Diemberger, H. (2007). *When a woman becomes a religious dynasty: The Samding Dorje Phagmo of Tibet.* New York: Columbia University Press.

Durham University. (2015). *Nepal earthquake landslide locations, 30 June 2015,* Durham University Nepal Landslide Data Website. Retrieved from https://data.hdx.rwlabs.org/dataset/nepal-earthquake-landslide-locations-30-june-2015.

Foreign Aid Commitment Doubled this Fiscal. (2015, July 11). *The Himalayan Times.* Retrieved from https://thehimalayantimes.com/business/foreign-aid-commitment-doubled-this-fiscal/.

Foreign Hand. (2016, February 5–11). The stifling burden of hierarchy. *The Nepali Times*. Retrieved from http://nepalitimes.com/regular-columns/Moving-Target/stifling-burden-of-hierarchy-in-Nepali-politics,659#.VrYTKBtd0w1.facebook.

Fuquan, Y. (2004). The "Ancient Tea and Horse Caravan Road," the "Silk Road" of Southwest China, *Silk Road Journal*, 2(1). Retrieved from http://www.silkroadfoundation.org/newsletter/2004vol2num1/tea.htm.

Fürer-Haimendorf, C. V. (1975). *Himalayan traders: Life in highland Nepal*. London: John Murray.

Giri, S. (2015, May 23). Power demand in Valley halves post-quake. *Kathmandu Post*. Retrieved from http://kathmandupost.ekantipur.com/printedition/news/2015-05-22/power-demand-in-valley-halves-post-quake.html.

Gleba, M., Vanden Berghe, I., & Aldenderfer, M. (2016). Textile technology in Nepal in the 5th–7th centuries CE: The case of Samdzong. *STAR: Science & Technology of Archaeological Research, 2*(1), 25–35. doi:10.1080/20548923.2015.1110421.

Government of Nepal, Department of Roads (GoN/DoR). (2013–2014). *Strategic Road Network 2013–2014*. Retrieved from http://www.dor.gov.np/documents/6.%20Strategic%20Road%20Network%202013_14%20Map.pdf.

Government of Nepal Ministry of Science, Technology and Environment (GoN/MoSTE). (2015). *Nepal earthquake 2015: Rapid environmental assessment*. Kathmandu, Nepal: Ministry of Science, Technology and Environment.

Govt Mulling Transit Treaty with China, Says Gyawali. (2015, November 1). *The Himalayan Times*. Retrieved from https://thehimalayantimes.com/kathmandu/govt-mulling-transit-treaty-with-china-says-gyawali/.

Gurung, P. (2016, January). *Cover story: Battling it out on Kathmandu roads*. Republica. Retrieved from http://admin.myrepublica.com/the-week/story/36038/cover-story-battling-it-out-on-kathmandu-roads.html.

Harvey, P., & Knox, H. (2015). *Roads: An anthropology of infrastructure and expertise*. Ithaca, New York: Cornell University Press.

Haviland, C. (2015, September 19). Why is Nepal's new constitution controversial? *BBC News*. Retrieved from http://www.bbc.com/news/world-asia-34280015.

Hayes, G., Briggs, R., Barnhart, W., et al. (2015). Rapid characterization of the 2015 M_w 7.8 Gorkha, Nepal, earthquake sequence and its seismotectonic context. *Seismological Research Letters, 86*(6), 1557–1567.

von der Heide, S. (2012). Linking routes from the silk road through Nepal – The ancient passage through Mustang and its importance as a Buddhist cultural landscape. In *Proceedings of the Archi-Cultural Translations through the Silk Road 2nd International Conference*, Mukogawa Women's Univ., Nishinomiya, Japan, July 14–16, 2012. Retrieved from http://www.mukogawau.ac.jp/~iasu2012/pdf/iaSU2012_Proceedings_613.pdf.

India to link Kathmandu with Delhi, Kolkata by rail. Economic Times. Retrieved from http://economictimes.indiatimes.com/articleshow/57257220.cms?utm_source=contentofinterest&utm_medium=text&utm_campaign=cppst.

International Center for Integrated Mountain Development (ICIMOD). (2015). *Impact of Nepal earthquake 2015 on Langthang Valley*. Retrieved from http://www.icimod.org/?q=20085.

Kargel, J. S., et al. (2015a). Geomorphic and geologic controls of geohazards induced by Nepal's 2015 Gorkha earthquake. *Science*. doi:10.1126/science.aac8353. Retrieved from sciencemag.org/content/early/recent.

Kargel, J. S., et al. (2015b). Geomorphic and geologic controls of geohazards induced by Nepal's 2015 Gorkha earthquake. *Science, 351*(6269), 140–150. Retrieved from http://dx.doi.org/10.1126/science.aac8353.

Khanal, R. (2015, October 30). Govt to procure fuel from China. *The Kathmandu Post*. Retrieved from http://kathmandupost.ekantipur.com/news/2015-10-25/govt-to-procure-fuel-from-china.html.

Knörzer, K.-H. (2000). 3000 years of agriculture in a valley of the high Himalayas. *Vegetation History and Archaeobotany, 9*, 219–222.

Lanxing, H. (1992). New archaeological findings in Tibet. *Beijing Review*, Updated: May 6, 2008 No. 31, 1992. Retrieved from http://www.bjreview.com.cn/special/tibet/txt/2008-05/06/content_115066.htm.

Lin, K. (2011). Le développement du réseau routier en chine: Inconéequences et inégalitiés. *Revue Internationale de Politique Comparée, 18*(3), 151–179.

Liu, J., Dietz, T., Carpenter, S., Alberti, M., Folke, C., Moran, E., Pell, A., Deadman, P., Kratz, T., Lubchenco, J., Ostrom, E., Ouyang, Z., Provencher, W., Redman, C. L., Schneider, S. H., & Taylor, W. (2007). Complexity of coupled human and natural systems. *Science, 317*(5844), 1513–1516.

Lord, A. (2016). Citizens of a hydropower nation: Territory and agency at the frontiers of hydropower development in Nepal. *Economic Anthropology, 3*, 145–160.

Lord, A. (2014). Making a 'hydropower nation:' subjectivity, mobility, and work in the Nepalese hydroscape. *HIMALAYA: Journal of the Association for Nepal and Himalayan Studies, 34*, 111–121.

MacFarlane, A. M., Hodges, K. V., & Lux, D. (1992). A structural analysis of the main central thrust zone, langtang National Park, Central Nepal Himalaya. *Geological Society of America Bulletin, 104*(11), 1389–1402.

Maillart, E. (1966). Nepal: The China road. *Journal of The Royal Central Asian Society, 53*(2), 143–146. doi:10.1080/03068376608731945.

Manandhar, R. (2014, August 8). Black market thrives after roadblock. *Kathmandu Post.* Retrieved from http://www.ekantipur.com/the-kathmandu-post/2014/08/08/nation/black-market-thrives-after-roadblock/265921.html.

Murton, G., Lord, A., & Beazley, R. (2016). "a handshake across the himalayas:" Chinese investment, hydropower development, and state formation in Nepal. *Eurasian Economics and Geography.* doi:10.1080/15387216.2016.1236349.

National Aeronautics and Space Administration (NASA). (2015). *Nepal Landslides 2015.* Retrieved from http://eoimages.gsfc.nasa.gov/images/imagerecords/87000/87172/nepal_2015_landslides_zoom_lrg.jpg.

Nepal Army. (2010). *Road projects of the Nepalese army.* Retrieved from http://www.nepalarmy.mil.np/bpd.php.

Panday, R. (1999). *Nepal's failed development: Reflections on the mission and the maladies.* Kathmandu: Nepal South Asia Centre.

Pokhrel, S. (2015, December 4). My voice: Road construction in Nepal. *Republica.* Retrieved from http://admin.myrepublica.com/lifestyle/story/32229/my-voice-road-construction-in-nepal.html.

Pokharel, K. (2015, November 26). The two-month blockade of Nepal explained. *The Wall Street Journal.* Retrieved from http://blogs.wsj.com/indiarealtime/2015/11/26/the-two-month-blockade-of-nepal-explained/.

Pradhani, K. (2015, October 29). Nepal finds China oil cheaper than India's. *The Times of India.* Retrieved from http://timesofindia.indiatimes.com/india/Nepal-finds-China-oil-cheaper-than-Indias/articleshow/49574590.cms.

Prasain, S., & Khanal, R. (2015, October 29). Nepal inks historic oil agreement with China. *The Kathmandu Post.* Retrieved from http://kathmandupost.ekantipur.com/printedition/news/2015-10-29/nepal-inks-historic-oil-agreement-with-china.html.

Puwar, N. (2004). *Space invaders: Race, gender and bodies out of place.* New York: Berg.

Qiu, J. (2015). Nepal earthquake caused fewer landslides than feared Satellite images show some mountains unscathed. *Nature–News,* Naature.com. Retrieved from http://www.nature.com/news/nepal-earthquake-caused-fewer-landslides-than-feared-1.19038.

Rai, O. (2016, June 10). Tapping Nepal's power potential. *China Daily Asia, Asia Weekly.* Retrieved from http://epaper.chinadailyasia.com/asia-weekly/article-8106.html.

Rana, P., & Karmacharya, B. (2014). *A connectivity-driven development strategy for Nepal: From a landlocked to a land-linked state.* ADBI Working Paper 498. Tokyo: Asian Development Bank Institute. Retrieved from http://www.adbi.org/working-paper/2014/09/08/6409.connectivity.dev.strategy.nepal/.

Rankin, K. (2004). *The cultural politics of markets: Economic liberalization and social change in Nepal.* London: Pluto Books.

Rathaur, H. (2015, November 3). Having brought in fuel, drivers demand Rasuwa road upgrade. *The Kathmandu Post*. Retrieved from http://epaper.ekantipur.com/epaper/the-kathmandu-post/2015-11-03/2015-11-03.pdf.

Regmi, M. (1970). *An official Nepali account of the Nepal-China war*, Regmi Research Series, Year 2, No. 8, (p. 177). Cornell University Library. (1969). Regmi research series. Kathmandu, Nepal: Regmi Research Ltd.

Regmi, A., Yoshida, K., Pourghasemi, H., et al. (2014). Landslide susceptibility mapping along Bhalubang—Shiwapur area of mid-western Nepal using frequency ratio and conditional probability models. *Journal of Mountain Science, 11*(5), 1266. doi:10.1007/s11629-013-2847-6.

Rigby, J. (2016, April 25). Nepal earthquake anniversary: One year on, not one home rebuilt by government. *The Telegraph*, telegraph.com. Retrieved from http://www.telegraph.co.uk/news/2016/04/25/nepal-earthquake-anniversary-one-year-on-not-one-home-rebuilt-by/.

Robb, E. S. M., et al. (2015). Geotechnical effects of the 2015 magnitude 7.8 Gorkha, Nepal, earthquake and aftershocks. *Seismological Research Letters, 86*(6), 1514–1523.

Rogers, C. (2004). *Secrets of Manang: The story behind the phenomenal rise of Nepal's famed business community*. Kathmandu: Mandala.

Samiti, R. (2015a, November 8). 27 Fuel tankers enter Nepal from China today. *The Himalayan Times*. Retrieved from https://thehimalayantimes.com/business/27-fuel-tankers-enter-nepal-from-china-today/.

Samiti, R. (2015b, November 1). Fuel will be imported via northern border: Maskey. *The Himalayan Times*. Retrieved from https://thehimalayantimes.com/nepal/fuel-will-be-imported-via-northern-border-maskey/.

Shakya, M. (2009). *Risk, vulnerability and tourism in developing countries: The case of Nepal*. Bochum Studies in International Development 56. Berlin.

Shakya, S. (2016, February 2). Reverse gear: Global advisory firms are now considering putting Nepal on a watch list for investments. *The Kathmandu Post*. Retrieved from http://kathmandu-post.ekantipur.com/news/2016-02-02/reverse-gear.html.

Shrestha, A. (2001). *Bleeding mountains of Nepal a story of corruption, greed, misuse of power and resources*. Kathmandu: Ekta Books.

Shrestha, A. (2014, July 31). China plans rail link with Nepal. *Kathmandu Post*. Retrieved from http://www.ekantipur.com/2014/07/31/business/china-plans-rail-link-with-nepal/392951.htm.

Shrestha, D. (2015a, December 18–24). The wild west: Rasuwa was projected as an alternative trade route to China, but is plagued by delays and corruption. *The Nepali Times*. Retrieved from http://nepalitimes.com/article/nation/delays-and-corruption-in-Rasuwa,2762.

Shrestha, H. (2015b, March 29). Xi announces Rs 14b in new aid. *Nepal Mountain News*. Retrieved from http://www.nepalmountainnews.com/cms/2015/03/29/xi-announces-rs-14b-in-new-aid/.

Shrestha, P. (2015c, November 6). 7 new routes to be used: Besides 2 in use, Nepal and China agree to upgrade passes being used by locals for trade. *The Kathmandu Post*. Retrieved from http://kathmandupost.ekantipur.com/news/2015-11-06/7-new-rou tes-to-be-used.html.

Shrestha, P., Lord, A., Mukherji, A., Shrestha, R. K., Yadav, L., Rai, N. (2016). *Benefit sharing and sustainable hydropower: Lessons from Nepal*. ICIMOD Research Report 2016/2. Kathmandu, Nepal.

Sigley, G. (2012). The ancient tea horse road: The politics of cultural heritage in Southwest China, *China Heritage Quarterly*, No. 29, March 2012. Retrieved from http://www.chinaheritagequarterly.org/features.php?searchterm=029_sigley.inc&issue=029.

Sijapati, B., Baniya, J., Bhandari, A., et al. (2015). *Migration and resilience: Experiences from the Nepal 2015 earthquake, research paper VIII*. Kathmandu: Centre for the Study of Labor and Mobility. Retrieved from http://www.ilo.org/wcmsp5/groups/public/---asia/---ro-bangkok/---ilo-kathmandu/documents/publication/wcms_379082.pdf.

Snellinger, A. (2013). *Kathmandu becoming a (Post) modern city: Road Expansion Project*. Retrieved from http://www.amandasnellinger.com/kathmandu-becoming-a-post-modern-city-road-expansion-project/.

Thapa, K. (2015a, May 30). Rasuwa starts relocating quake displaced people. *The Kathmandu Post*. Retrieved from http://kathmandupost.ekantipur.com/news/2015-08-06/rasuwa-starts-relocating-quake-displaced-people.html.

Thapa, K. (2015b, October 31). Tankers in north border to take fuel from China. *The Kathmandu Post*. Retrieved from http://kathmandupost.ekantipur.com/news/2015-10-31/tankers-in-north-border-to-take-fuel-from-china.html.

Thapa, K. (2015c, November 2). 12 tankers arrive from Kerung: Bring in 80,000 litre petrol in first lot, *The Kathmandu Post*. Retrieved from http://kathmandupost.ekantipur.com/news/2015-11-02/12-tankers-arrive-from-kerung.html.

Tiwari, A. (2016, June) *Biratnagar developmental projects in limbo*. Republica. Retrieved from http://admin.myrepublica.com/society/story/44505/biratnagar-developmental-projects-in-limbo.html.

United Nations Development Program (UNDP). (2006). *Developing rural transport and infrastructure. Millennium development goals needs assessment for Nepal*. Retrieved from http://www.undp.org.np/publication/html/mdg_NAN/Chapter_9.pdf.

United States Agency International Development (USAID). (2015a, June 25). *Nepal earthquake fact sheet #21*, Fiscal Year (FY) 2015. Retrieved from http://www.usaid.gov/what-we-do/working-crises-and-conflict/responding-times-crisis/where-we-work.

United States Agency International Development (USAID). (2015b, September 2). *Nepal earthquake fact sheet #23*, Fiscal Year (FY) 2015. Retrieved from http://www.usaid.gov/what-we-do/working-crises-and-conflict/responding-times-crisis/where-we-work.

United States Agency International Development (USAID). (2015c, December 23). *Nepal earthquake fact sheet #1*, Fiscal Year (FY) 2016. Retrieved from http://www.usaid.gov/what-we-do/working-crises-and-conflict/responding-times-crisis/where-we-work.

Uteng, T. (2011). *Gender and mobility in the developing World*. World Development Report 2012 gender equality and development background Paper. Retrieved from http://siteresources.worldbank.org/INTWDR2012/Resources/7778105-1299699968583/7786210-1322671773271/uteng.pdf.

Van Spengen, W. (2000). *Tibetan border worlds: A geohistorical analysis of trade and traders*. New York: Kegan Paul International.

Vehicles Carrying Chinese Imports to Nepal Arrive in Kyirong. (2015, November 6). *Setopati*. Retrieved from http://setopati.net/politics/10091/Vehicles-carrying-Chinese-imports-to-Nepal-arrive-in-Kerung/.

Walker, B. (2015, April 29). *Chinese dam workers stranded after Nepal Quake, thethirdpole.net*. Retrieved from http://www.thethirdpole.net/2015/04/29/chinese-dam-workers-stranded-after-nepal-quake/.

WB Okays $55m Loan for Roads, Bridges. (2016, August 27). *The Kathmandu Post*. Retrieved from http://kathmandupost.ekantipur.com/news/2016-08-27/wb-okays-55m-loan-for-roads-bridges.html.

World Bank (WB). (2005). *Nepal: North south transport corridor options*, International Development Association Assistance Strategy Note, March 1, 2005. Retrieved from http://siteresources.worldbank.org/INTSARREGTOPTRANSPORT/34004316-1111699655514/20690624/NorthSouthCorridor03105.pdf.

Yan, C. (2011, January 18). Road-building rage to leave U.S. in dust, China real time report. *The Wall Street Journal*. Retrieved from http://blogs.wsj.com/chinarealtime/2011/01/18/road-building-rage-to-leave-us-in-dust/.

Yang, B. (2009). *Between winds and clouds: The making of Yunnan (second century BCE to twentieth century CE)*. New York: Columbia University Press.

Yeh, E. (2013). *Taming Tibet*. Ithaca: Cornell University Press.

Index

© The Author(s) 2017
R.E. Beazley, J.P. Lassoie, *Himalayan Mobilities*, SpringerBriefs
in Environmental Science, DOI 10.1007/978-3-319-55757-1